Evolution and the Victorians

Frontispiece: Charles Darwin, 'Notebook B' (1837–38), f. 36.

Evolution and the Victorians

Science, Culture and Politics in Darwin's Britain

JONATHAN CONLIN

BLOOMSBURY
LONDON • NEW DELHI • NEW YORK • SYDNEY

Bloomsbury Academic
An imprint of Bloomsbury Publishing Plc

50 Bedford Square
London
WC1B 3DP
UK

1385 Broadway
New York
NY 10018
USA

www.bloomsbury.com

Bloomsbury is a registered trade mark of Bloomsbury Publishing Plc

First published 2014

British Library Cataloguing-in-Publication Data
A catalogue record for this book is available from the British Library.

ISBN: HB: 978-1-4411-3609-1
PB: 978-1-4411-3090-7
ePDF: 978-1-4411-8752-9
ePub: 978-1-4411-2613-9

Library of Congress Cataloging-in-Publication Data
A catalog record for this book is available from the Library of Congress.

Typeset by Deanta Global Publishing Services, Chennai, India
Printed and bound in India

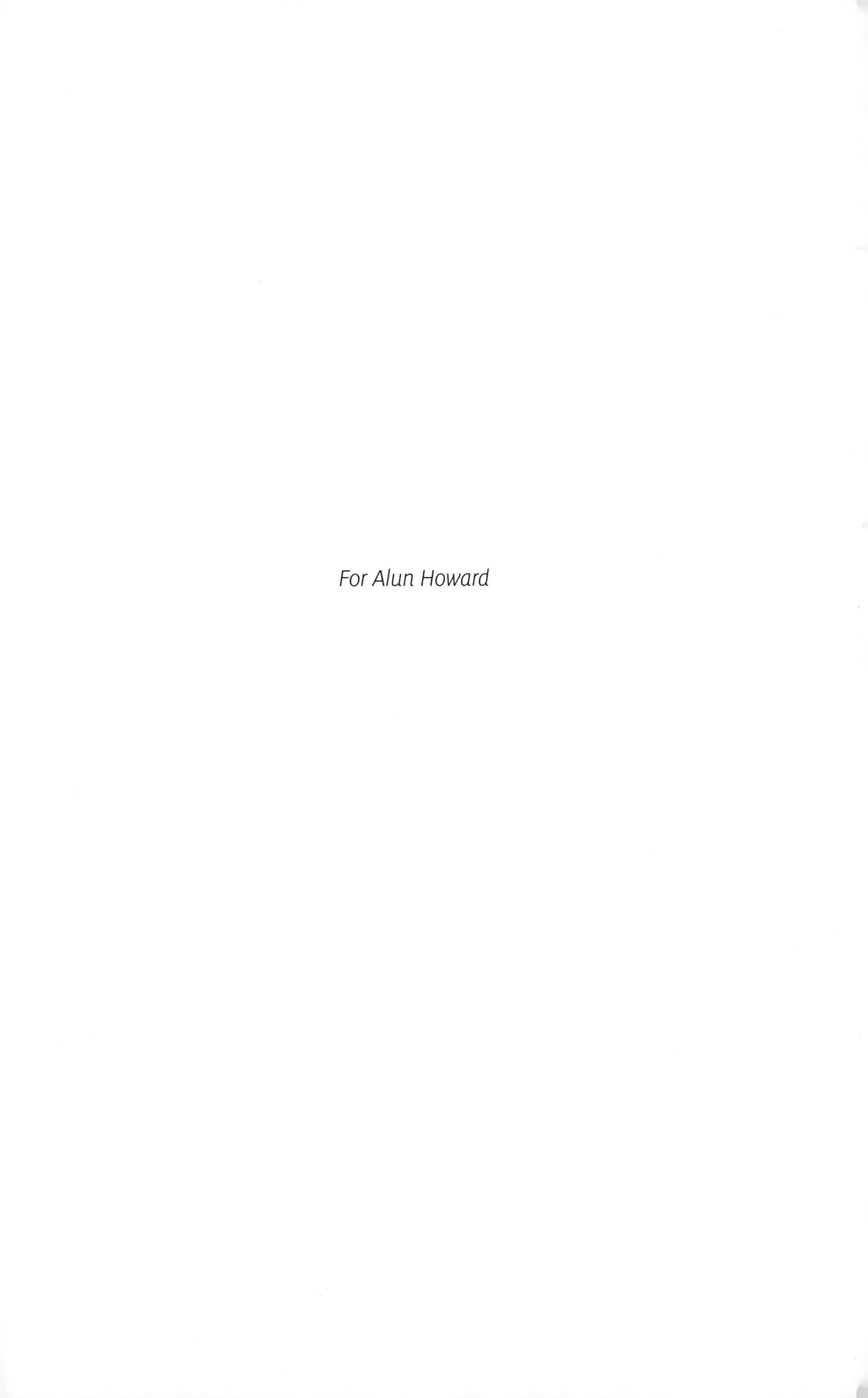

For Alun Howard

CONTENTS

ACKNOWLEDGEMENTS

This book is based on 'The Longest Discovery: Darwin and the Politics of Evolution,' an undergraduate history course I taught at the University of Southampton from 2006 until 2011. I would like to thank my former students for making this experience so enjoyable, particularly those who went on to research and write undergraduate and postgraduate dissertations on related topics: Laura Mainwaring, Curtis McGlinchey, Alex Reynolds and particularly Ahren Lester. Like them, I was coming to the history of science from outside, from a background in British political and cultural history. If this book serves to make that transition easier for others, by providing a helpful introduction to a specialist field of historical enquiry, then that is due to the generosity with which several leading historians of science gave of their time and expertise. My greatest debts are to Pietro Corsi, Bernard Lightman, David Stack and one anonymous reader for their detailed responses to drafts. At Southampton the biologists John Allen and Lex Kraaijeveld took the time to explain current trends in evolutionary science to my students and I, while at Cambridge Rebecca Kilner kindly agreed to proofread my attempt to explain 'evo devo' in layman's terms. I am grateful to Anne Barrett and Judith Magee for granting my students and I access to the notebooks and other papers of Huxley and Wallace held at Imperial College and the Natural History Museum. At Bloomsbury Academic I would like to thank Rhodri Mogford for inviting me to write the book.

LIST OF ILLUSTRATIONS

TIMELINE

Science

1802	William Paley, *Natural Theology*
1807	Geological Society of London founded
1813	Robert Jameson publishes Cuvier's *Discours préliminaire* in English.
1826	Society for the Diffusion of Useful Knowledge founded
1831	British Association for the Advancement of Science (BAAS) founded
1833	First of Bridgewater Treatises published
1836	Richard Owen appointed Hunterian Professor
1844	[Chambers], *Vestiges of the Natural History of Creation*
1854	Huxley appointed Professor of Natural History at the School of Mines
1856	First recognized fossil of a Neanderthal found in Germany
1859	Darwin, *On the Origin of Species*
1860	Wilberforce/Huxley "debate" at BAAS in Oxford
1861	Fossil of *Archaeopteryx* ("first bird") skeleton found in Germany
1863	Kingsley, *The Water-Babies*
1864	X Club founded
1869	Huxley appointed president of the BAAS
1870	Physicist William Thomson proposes radically shorter history for earth
1871	Darwin, *The Descent of Man*; Anthropological Institute established
1874	John Tyndall's Belfast Address to the BAAS
1881	Natural History Museum opens
1882	Society for Psychical Research founded
1894	Salisbury's address to BAAS pronounces evolution "not proven"

The Wider World

1776 American Revolution

1789 French Revolution

1807 Slave trade abolished

1815 Battle of Waterloo/end of Napoleonic Wars
1817 Habeas Corpus suspended (owing to fears of reform/sedition)

1829 Catholic Emancipation (removal of restrictions on Roman Catholics)
1830 Liverpool and Manchester Railway opens (first in UK)

1832 Great Reform Act
1833 Slavery abolished in British Empire
1834 New Poor Law
1838 People's Charter drawn up (basis of Chartist movement)

1845 John Henry Newman converts to Roman Catholicism
1846 Corn Laws repealed
1848 Revolutions in France, German and Italian states

1851 Great Exhibition
1857 Indian Rebellion
1858 *SS Great Eastern* launched (largest steamship in the world)

1865 Morant Bay Rebellion
1866 First transatlantic telegraph cable laid
1867 Second Reform Act
1868 Gladstone becomes prime minister for the first time
1870 Forster's Education Act

1886 Gladstone fails to pass Home Rule for Ireland, splits Liberals
1887 Bloody Sunday Riot in London

1895 Oscar Wilde imprisoned after 'gross indecency' conviction
1899 Second Boer War (–1902)

1903 Women's Social and Political Union founded
1905 Aliens Act (UK's first restrictions on immigration)

1911 National Insurance Act lays foundation of welfare state
1914 World War I breaks out

A NOTE ON CURRENCY

Where contemporary equivalents are given for Victorian prices and salaries the figures are for 2005 purchasing power parity, using the historic currency converter on the National Archives' website, for the round date closest in time to the year in question (I use the 1860 rate for 1865 prices, 'rounding down' for any year ending in 5).
http://www.nationalarchives.gov.uk/currency

Introduction: ‘I think’

FIGURE 1 *Maull and Polyblank, Charles Darwin, c. 1855. Even before the publication of* On the Origin of Species *Darwin’s* Beagle *voyage and associated publications had earned him a place in the* Literary and Scientific Portrait Club, *a ‘virtual club’ consisting of photographs of eminent Victorians. Only the eyes and brow hint at the ‘mental rioting’ beneath.*

At some point between July 1837 and January 1838 Charles Darwin wrote these words at the top of page 36 of 'Notebook B', adding a small sketch [Frontispiece] of a tree with various letters on its branches below. The palm-sized notebook was one of a series that Darwin had on the go at this point, each of them devoted to one of his scientific interests. B and M were devoted to transmutation, to the theory that species had not remained fixed since Creation, but had changed. As the scribblings around the sketch indicate, observations made on a voyage around the world on *HMS Beagle*, as well as wide reading, had led Darwin to speculate that a species might become extinct and yet leave behind several new species as its descendants. The new species would share a number of features. For centuries humans had used such similarities to classify living things into tidy groups. The tidiness of this classification was believed to reveal the tidiness of God's Creation, in which there was a place for everything, and everything had its place. What if, Darwin speculated, those features were marks of inheritance, rather than a divine plan?

The sketch was the first time Darwin had drawn the tree diagram which has since become synonymous with evolution. Another one would appear in his 1859 book *On The Origin Of Species*, showing the evolutionary links between wild and domesticated pigeons. One hundred and fifty years later the 'I think' sketch could be found on postcards, t-shirts, even as a tattoo. The world was celebrating the bicentenary of Darwin's birth and the 150th anniversary of his book, *On the Origin of Species*. Darwin's theory of evolution by natural selection was hailed as one of the greatest steps in mankind's ongoing project to know and understand the universe. Darwin had been born into a world lovingly created around the needs of one species, *Homo sapiens*, created by God in His own likeness thousands of years before. Darwin and his fellow Victorians lived to see humanity and God banished to the edge of that universe: the former reduced to the level of an improved ape, the latter missing, presumed dead. The universe was no longer 'our' universe, constructed around our needs. Darwin was under no illusions about the upset his discovery would cause. 'It is like confessing a murder', he wrote to a close friend.[1]

Few of the readers of this book will have been born into or grown up in a Creator-focused universe, or see their lives in terms of a relationship with a Supernatural Being. Such is the power of evolution as an idea, however, that it cannot help but confront us, all of us, 'religious,' agnostic or atheist, with unsettling questions. What does it mean to be human? Are we masters of our own destiny, able to make a free choice among real alternatives? Is our destiny in the hands of our genes, which really pull the strings? Can we resist the commands of these genes, these 'selfish replicators', as Richard Dawkins calls them in his famous book, *The Selfish Gene*?

[1]Charles Darwin to Joseph Hooker, 11 January 1844. Darwin Correspondence Database, http://www.darwinproject.ac.uk 729 (accessed on 2 May 2013).

Of course, the Victorians did not see evolution in genetic terms, as the gene had yet to be identified. Yet the idea of evolution inspired questions similar to those it confronts us with today. For the Victorian biologist Thomas Henry Huxley evolution posed 'The question of questions for mankind':

> Whence our race has come; what are the limits of our power over nature, and of nature's power over us; to what good we are tending; are the problems which present themselves anew and with undiminished interest to every man born into the world.[2]

For all their impressiveness, the television programmes, exhibitions, books and websites broadcast, organized, written and launched in 2009 represented just one chapter in the story of how humans have answered these questions. This book is an introduction for historians interested in how Darwin came up with his theory in the early nineteenth century and how he came to publish, defend and refine it. This process itself represents another chapter (if an important one) in the same story.

Darwin did not go it alone, however. Without the help of family, mentors, fellow travellers, servants and other gentlemen of science he would not have been able to collect the wide range of observations which allowed him to question accepted theories, advance new ones and test them. A shy, retiring man plagued by illness, after 1842 Darwin hardly left his secluded home in Down, Kent, except for medical treatment. In addition to notes and doodles like the 'I think' sketch Darwin wrote longer 'sketches' of his theory in 1842 and 1844. But he did not publish them, leaving instructions for his wife on what to do with them after his death. Had it not been for the independent discovery of the theory in 1858 by another British naturalist, Alfred Russel Wallace, the theory might have remained locked up in Darwin's desk. If one were holding casting calls for the role of scientific hero, a pioneer who would bring an outrageous new truth to a sceptical world, Darwin would not have been the natural selection.

Wallace's letter to Darwin outlining his own theory was a striking incident of parallel discovery, one Darwin himself found uncanny as well as surprising. On the advice of several friends Darwin and Wallace's new theory was formally presented to the scientific community at a meeting of the Linnean Society in London, held on the evening of 1 July 1858. Both men were absent. Wallace had reasonably, if modestly, acknowledged that Darwin should be credited with the discovery, in view of Darwin's 20-odd years of covert labour on it. The response of the audience at the Linnean Society was muted. Summing up the year in December 1858 Linnean Society president Thomas Bell remarked that it had not been marked 'by any of those striking discoveries which at once revolutionize, so to speak, [our]

---

[2]Thomas Henry Huxley, *Evidence as to Man's Place in Nature* (New York: Appleton, 1863), p. 71.

department of science'.[3] Only when copies of the hastily written *On the Origin of Species* appeared in November 1859 did Darwin's discovery begin to get the kind of response we might expect.

The following year there was a great confrontation over the new theory at Oxford, with the Anglican Bishop of Oxford, Samuel Wilberforce, on the one side and Darwin's ally, Huxley, on the other. Taunts were thrown. Someone held a large Bible in the air, loudly insisting on the authority of scripture. A lady fainted. Students cheered. Evolution and the Established Church were at war. Admittedly, Darwin was absent, again. But after 20 years of silence, after the hesitations, the doubts and the incessant scribbling in notebooks – which make it so hard for us to pin down the *when* of this greatest of all discoveries – one cannot help feeling that this was more like it. This was dramatic, this was History-with-a-capital-H, the kind directors of history documentaries love to reconstruct.

Although the teaching of history is no longer structured around the rote learning of dates, historians remain determined to identify when changes occurred, to locate turning points in time, forks in the path of history as tidy as the forks of Darwin's tree. In 1655 the Vice-Chancellor of Cambridge University, John Lightfoot, used familiar historical methods, for example, to work out the age of the world. Jesus Christ's career as a freedom-fighter in Roman-era Judaea gave Lightfoot a point from which to start working backwards, using the family pedigree provided in Genesis and the Gospels to trace Jesus' ancestry back through David, Abraham and Noah to Adam, the first human, created on the sixth day. The world was created, Lightfoot concluded, at 9 a.m. on Sunday, 23 October 4004 BCE.

The anecdote is humorous, for two reasons. First because it has since been proved to be wildly wrong, by several orders of magnitude. Current estimates put the age of the universe at between 4 and 5 *billion* years. But it is also ludicrous because of its pretensions to accuracy. Though the Big Bang theory is widely accepted, we know too much about the limitations of our current understanding to presume to give this supreme event a historical date. Indeed, many physicists might see the question of when exactly it happened as unknowable, or simply pointless, given our understanding of how time itself can be bent and folded. There are more urgent questions for science to tackle.

Yet when it comes to the *history* of science the general public resembles Lightfoot in seeking neatly defined events. The ancient Greek philosopher Archimedes is supposed to have jumped into his bath, noticed that the water level rose and shouted 'Eureka' ('I have found it!'), delighted at realizing he had discovered a way of measuring the volume of irregularly shaped objects. He is supposed to have been so eager to share his discovery that he ran through the streets of his home town, Syracuse, forgetting in his haste to put

---

[3]A. T. Gage and W. T. Stearn, *A Bicentenary History of the Linnean Society of London* (London: Academic Press, 1988), p. 57.

on any clothes. Isaac Newton's supposed discovery of gravity in seventeenth-century Cambridge has also been attributed to a 'eureka moment', when an apple dropped on his head. The cartoon lightbulb going on above someone's head has the same force, suggesting an instantaneous shift from ignorance to knowledge. Discovery is a 'revelation', it 'comes like a bolt from the blue': in telling such stories we cannot help, it seems, but use a language that Lightfoot or any other believer in supernatural influence might use.

In this book we will try to avoid the tendency to see 1859 as 'year zero,' to understand everything that came after in terms of a struggle between 'Darwinist' and 'Anti-Darwinist' camps. A closer look at the debates that raged in nineteenth-century Britain reveals the limitations of our usual way of understanding how knowledge is advanced: as a series of moments in which scholar-recluses emerge blinking into the glare of public acceptance. In the case of Darwin many refused to accept. Others accepted, but changed the theory in the process. Darwin must have been tempted to beg for protection from those claiming to be his friends and supporters.

'Evolution' and 'the theory of natural selection' were not synonymous. Natural selection as proposed in *The Origin* was just one mechanism by which species change was explained. In addition to tweaking his own version of natural selection Darwin came up with another mechanism: sexual selection. Other gentlemen of science proposed other mechanisms. And these mechanisms were not necessarily alternatives to one another. Darwin and his colleagues were caught up in a long discussion over how much weight to give different mechanisms in explaining transmutation. Few were purists, that is, few felt that one mechanism alone could explain evolution across all time and for all species, alive and extinct. The mechanism could be different for the same species at different points in its course. Darwin only used the word 'evolved' once in *The Origin* ('evolution' never appeared at all), fearing the word's implications. Many who came before and after Darwin saw 'evolution' (which literally means 'unfolding') as the gradual unravelling or revealing of a divine plan. Even those who did not believe in a divine Creator often assumed that evolutionary mechanisms like natural selection were engines moving along a preordained, progressive track (e.g. from 'lower' to 'higher' forms of life). The vast majority of Victorians were 'teleological' in that they saw evolution as a process directed towards a certain goal.

Then as now, science was not a distinct realm of human activity inhabited by otherworldly, disembodied brains separated from the hopes and fears of the rest of society. Patronage, ideology, belief, rhetoric all played a part in the painfully drawn-out emergence of the theory and in debate on its implications for men and women, employers and employees, for the state and for its citizens. In considering the history of Victorian science we therefore need to keep other, equally important strands of British history in view: 'reform', urbanization, industrialization, imperial expansion, secularization, the extension of the franchise and the emergence of state welfare and education policy are some of these. These strands are so interwoven with the one considered here that 'placing science in its historic context' is not enough. Nineteenth-century views of free trade and economic development helped inspire Darwin's discovery, and Darwin's ideas in turn influenced how Britons viewed inequality, unemployment and industrialization.

For us historians, this is challenging, but also reassuring. Were it not the case there would be little point to studying the history of this or any other chapter in the history of science. Yet many people do see discovery in ahistorical terms: as a guided missile, homing in on truth. It is possible, they concede, temporarily to knock this missile off course, but it always 'locks on' again afterwards. Seen in this way, the history of science is either a relentless 'march of mind' (as the great historian, T. B. Macaulay, might have called it), or a history of tiresome obstruction and delay, of conservative institutions (the Roman Catholic church, say) or wicked individuals standing in the way for their own selfish reasons. At its most severe we might see the Establishment as carrying out a tactical retreat, as double-minded: continuing to advocate a set of beliefs that they actually did not believe in, 'deep-down'.

This so-called positivist approach to the history of science is often opposed to an 'environmentalist' one, which places greater emphasis on the role of the society and the individual in shaping events. Though very different in other respects, like the positivist approach it invites us to query the familiar view of scientific heroes as geniuses. If 'the truth will out' how much credit can Darwin and Wallace claim? Somebody else would have come up with the idea eventually, if they hadn't. If natural selection was 'in the air' or a 'product of its time', how far is it theirs to claim? For historians, the latter question is nothing new. All history makes us think about how we as individuals relate to our society. The historian of science John Henry identifies another benefit of our study, one particular to the history of science. It can 'help us to understand the nature of science itself: its methods, its "logic of discovery," how consensus is reached, how controversies are settled, whether (or to what extent) its institutional forms directly affect its doctrines and practices'.[4]

Achieving a consensus on the relationship between the activity we call 'science' and society is difficult, and this book cannot pretend to offer more than some hints for further consideration and discussion. The same reservations apply to this book's approach to the truth claims made by philosophers, churchmen and scientists, to the authority claimed by this or that theory about the universe's origin and development. Before we begin we need to acknowledge the role belief (of a non-religious kind) plays in science. But first it may be helpful to have an explanation of how Darwin and his contemporaries developed their theories, in response to a crisis in

---

[4]John Henry, 'Ideology, inevitability and the scientific revolution,' *Isis* 99 (2008): 552–9 (559).

the system by which naturalists visualized the diversity of natural forms. The next two sections introduce the Linnaean system of classification as well as the theory of natural selection and show how we have refined the theory since.

Darwin's problem with species

According to Genesis God created man on the sixth day and defined man's relationship to everything God had created over the previous 5 days of Creation. Plants were to yield food to man (humans were vegetarian), while humans were to have 'dominion' over all animals. God then has all the animals and birds line up and presents them to Adam 'to see what he would call them: and whatsoever Adam called every living creature, that was the name thereof' (Genesis 2:19-20). Authority ('dominion') and naming (taxonomy) are intimately related. Whether the thing we are naming is a beetle, a people or a language, to name is to stake a claim to a kind of ownership. When the astronomer William Herschel named the new planet (Uranus) he discovered in 1781 'the Georgian star' after King George III he was staking a claim for British astronomy *and* making a barefaced play for more research funding. Putting the things we name into groups is the next step by which we impose order on the unruly diversity of life. It was easy to fit this account of dominion with the concept of a 'great chain of being' (originally developed by the ancient Greek philosopher Plato) and see all forms of life as arranged hierarchically, step by step, from inanimate rocks up through humans and angels to God.

The resulting names and orders seem self-evident to us, can seem 'natural', that is, part of the thing, rather than something we attribute to it. We call a fish 'fish' because, well, it *is* a fish. There seems to be something equally 'obvious' about the ways in which we group things into classes: all swimming things go over there, flying things over here. They don't just 'go' there, they really seem to 'belong' there. Darwin and his contemporaries used a system of binomial classification invented in the previous century by a Swede, Charles Linnaeus (also known as Carl von Linné 1707–78). Linnaeus began by defining a species as a group of living things which share certain physical features (a morphology). Similar species were gathered in a group called a genus; similar genuses were gathered together into orders, and so on up through various levels of a hierarchy, up to classes. Although this system remains in use, and continues to use the same dead languages (Latin and Greek), we have added four more 'ranks' to this hierarchy: families, kingdoms, domains and phyla.

As the adjective 'binomial' indicates, Linnaeus' system gave each creature a name made up of both its genus and its species. Humans or *Homo sapiens* are thus species *sapiens* of the genus *Homo*. Although Linnaeus presented this system more like a table, we are used to representing this classificatory order as a series of pyramids: the five classes at the apex, thousands of species at the bottom, with each level of classification 'nesting' in the one above. Sometimes we draw lines, making the pyramid look like an upside-down tree. Ever since Linnaeus naturalists have been hard at work within his taxonomic architecture. The task they assigned themselves involved assigning newly discovered forms of life to the class, order and genus they 'belonged' to, then coming up with a species name. This species name could be based on a trait, as with *Homo sapiens* (*sapiens* Latin for 'wise'); on geography, as with *Homo floriensis* (named after the island of Flores, where the bones of this extinct 'Hobbit' were found); or on somebody's name, as with the spider *Aptostichus angelinajolieae* (named after an actor, Angelina Jolie). Occasionally a species might be reassigned, one genus or class might be carved up into several different ones, but there was a place in the Linnaean system for every form of life.

For Darwin and his contemporaries, the neatness and elegance of this structure were not accidental. They were a direct reflection of the neatness and elegance of the supremely intelligent mind who had thought it all up in the first place, the mind of the Creator. If there were gaps in the puzzle, or if certain pieces just did not fit, that was because our mortal minds were not as intelligent as God's, or perhaps because the 'missing pieces' had been lost in one of the mass extinctions orchestrated by God. Noah's Flood was the most familiar of these divine catastrophes. Though species in the same genus inevitably bore a strong resemblance to each other, a firm barrier to species change remained. Members of two different species could not breed with each other and produce fertile offspring. To use today's scientific jargon, they were intersterile. The concept of a variety, a level of classification 'below' species, allowed a certain room to accommodate troublesome creatures which did not quite fit in their assigned places, without having to give them full species recognition. Other organisms which might have been seen as 'intermediate forms' (missing links, as it were) could be dismissed as 'sports of nature': freaks, in other words. By such means one could preserve the gaps necessary to maintain a tidy separation between orders and classes.

Though members of the same genus or order might be connected by a set of morphological relationships, there was no sense at all that these were a shared inheritance from a common ancestor. This was a static, fixed order. Extinct species were certainly recovered from underground fossil beds, but each and every one of them had been created by God from nothing, at the same time as similar creatures that were still living. It is important to emphasize this, as it is hard for us, now, to look at a Linnaean pyramid without mentally flipping it upside down and turning it into a family tree, into Darwin's 'I think' diagram. In turning this neat, elegant pyramid upside down Darwin turned a fixed, complete, perfect plan into a shifting, incomplete, messy narrative. The 'work' of assigning forms of life to Linnaeaus' system went from being descriptive to being historical.

In classifying a life form we are not only placing it among all other living things, but are also giving it a past, a pedigree or lineage.

Darwin and his contemporaries were led to this in the early nineteenth century partly by French theories of species change (transmutation) advanced by Lamarck, discussed in Chapter 2, and partly because expeditions to far-flung regions were bringing back forms of life (both alive and extinct) which did not seem to fit the plan. These 'doubtful forms' (as Darwin called them) were making it hard to achieve that consensus necessary to science, by its very nature a collaborative project. One naturalist's variety was another naturalist's species. The species to which he (and it was always a he) assigned creatures was different, depending on whether he distinguished them from one another by, say, their dentition (arrangement of teeth) or by other organs or structures.

Of course, when challenged, the naturalist could provide evidence that species were fixed, by pointing to the fact that these 'important' organs (e.g. teeth) did *not* vary, even if other, 'unimportant' ones did. Yet, as Darwin noted, this was a circular argument, 'for these same authors practically rank that character [i.e. that trait] as important (as some few naturalists have honestly confessed) which does not vary; and, under this point of view, no instance of an important part varying will ever be found.'[5] These 'systematists' were putting their system first, yet in trying to accommodate 'doubtful forms' they were having to add patches to the plan that were far from elegant. To save the plan, Darwin would have to destroy it.

As he wrote in *The Origin*, naturalists had been looking at nature 'as a savage looks at a ship, as at something wholly beyond his comprehension'. Darwin was thinking of the cannibals he had previously encountered in Tierra del Fuego, on the southern tip of South America. While the Fuegians had paid close and admiring attention to an unfamiliar colour of cloth or beads, Darwin observed, they showed no interest in the ship. The *Beagle* was a vessel made, like the Fuegians' canoes, of wood, but was much larger, sophisticated and capable of far longer voyages. Yet the Fuegians saw the ship as a natural phenomenon, not something contrived by humans to fulfil a purpose. In presenting his theory of transmutation Darwin proposed not only to fix the problems with the Linnaean system but also to make the work of assigning forms to their order and genus 'far more interesting'. Natural history would be enriched as we learned to 'contemplate every complex structure and instinct as the summing up of many contrivances, each useful to the possessor'.[6]

Darwin's theory of natural selection was a theory of speciation, that is, it proposed a mechanism by which one species changed into another, or into

---

[5]Charles Darwin, *On The Origin of Species By Means of Natural Selection* (Mineola, NY: Dover, 2006), p. 30.

[6]Ibid., p. 304.

several other species over long periods of time. To work, natural selection needed mutation, superfecundity and a changing environment. When forms reproduced sexually, the offspring were not perfect copies of their parents, Darwin noted. Small changes or mutations occurred. Some of these changes would be 'useful to the possessor', others harmful. Meanwhile Darwin noted that forms will reproduce exponentially. Forms could produce so many offspring so quickly (hence *super*fecundity) that they would soon reach the point at which there would no longer be room on the planet for them to stand up in. Obviously there were limiting factors or checks to population that were constantly working to keep population from reaching this point. Though the number of forms would increase exponentially, their food supply would not. They would therefore have to compete with one another for what food there was.

Superfecundity thus doomed many forms that lost out in this struggle to an early death. It would not take much of an advantage to make the difference between getting enough to live and starving. For those forms that were prey to other forms, an equally small advantage would make the difference between surviving to breed and being eaten before one had the chance. An old joke has two hikers meet an angry bear in the woods. Both stop, and one uses the time to get out a pair of running shoes and put them on. His companion, bewildered, asks if he thinks he can outrun a bear. 'Oh, I don't have to outrun the bear,' comes the reply, 'I only have to outrun you.'

As the phrase 'survival of the fittest' has it, members of the same species are constantly competing with one another. Those with beneficial mutations, with mutations that increase 'fitness' (better traction, in the bear example), will not only survive where those without die, but will also tend to be more successful when it comes to breeding. Beneficial mutations will become more widespread in the population as time passes, slowly changing the species' morphology. It is important not to see this as making the species a 'more highly evolved' one. There is a widespread misconception that evolution is about everything getting better, a process by which every form of life becomes faster (musclier?), tougher (armour-plated?), deadlier (sharper teeth?) and smarter (bigger brains?). By this theory humans are going to develop into super-beings with massive brains. In our confusion of 'evolution' with 'progress', in our teleological tendency, we are like the Victorians.

Avoiding unhelpful qualifiers such as 'more' or 'less' when speaking of evolution is hard. Darwin made a resolution not to use the adjectives 'high' and 'low' when writing about forms of life, and didn't keep it, sometimes writing of a tendency towards 'perfection' in the natural world. The fact remains that 'fitness' is judged not in universal or absolute terms, but with respect to the environment at a certain point in time. Darwin's discovery is often associated with the finches he encountered in the Galapagos Islands, off the coast of Ecuador, which provide good examples of how fitness is related to environment. All of the Galapagos Islands had been created by volcanic activity, but not all at the same time. Hence some were still an arid moonscape when Darwin visited, while others had had time to grow lush vegetation, from seeds that had been carried from the mainland by the wind. The former tended to have seed-bearing plants, the latter fruit-bearing trees and insects. A finch born with a slightly thicker beak counted as 'fitter' on an arid island, because wide, thick beaks are better at breaking open seeds. But a finch born with the same beak on a lush island would be *less* fit. Small insects are easier to pick up with a pair of tweezers than with a nutcracker, and so finches with thinner beaks would count as 'fit'. Over a vast period of time one species of finch had become several, each adapted to the climate and vegetation of a different island in the Galapagos chain.

Without a range of different environments and without a process of environmental change it would be hard to explain why there is as much diversity of life on this planet as there is. The driver of environmental change in the case of the Galapagos Islands was upheaval in the earth's crust, a function of the ongoing process by which tectonic plates are formed, move, slip and collide, causing earthquakes and undersea vents as well as volcanoes. Apart from his early collecting of beetles Darwin's first efforts as a gentleman of science were focused on geology, on that of the geologist Charles Lyell in particular. Over a vast scale of time rises in sea level along with other factors such as glaciation (Ice Ages) have stranded life, allowing a certain stability to appear on an island or continent. Falls in sea level have caused land bridges to appear, such as that which once linked Asia and North America, confronting life forms with new competition and new opportunities.

It isn't just the 'hard' stuff of shifting landmasses that counts. There are many other drivers of environmental change. Land masses may remain roughly where they are, but the climate may change, along with ocean and wind currents. Vegetation can change, creating new environments such as forest. There are extraterrestrial drivers. Asteroid impacts and shifts in the earth's orbit around the sun are some of these. By casting up a massive cloud of dust into the atmosphere the former may have brought on an impact winter sufficient to bring about radical changes in life on earth in a relatively short space of time. 'Environment' also includes other life forms that are themselves changing, and which may outcompete other species. Superfecundity is constantly pushing life to explore the new conditions and places uncovered or forged by environmental change. By specializing life forms can find a niche in which to thrive; the more specialized, the smaller the threat of competition. Mutation provides tools to colonize these niches successfully, at least until the environment changes, whereupon a cosy niche can become a trap. In a dynamic environment it doesn't always pay to be highly specialized. Having a mid-sized beak able to eat seeds *and* fruit may be a better strategy.

Terms like 'strategy' can create a misleading impression that the adaptations made by finches and other creatures are the result of a conscious choice made by the species themselves or by some higher power. The very term 'natural selection' implies a selector, as Darwin recognized. This is why Wallace was reluctant to use it. Yet the mutations are entirely random, and there is no will or mind at work in selecting among them. The process by which natural selection filters out the beneficial mutations from the rest is also phenomenally expensive in terms of the lives lost along the way. This would prove to be a major obstacle to public acceptance of the theory, along with a resistance to seeing a complex web of interlocking species as the product of eons of hit-and-miss, of what the astronomer John Herschel (William's son) called 'the law of higgledy-piggledy'.[7]

Evolution after Darwin

Darwin was uninterested in explaining how life originated and what produced the mutations. We have little progress to show in addressing the former, but the discovery of the gene has greatly advanced our understanding of the latter. For his part, Darwin advanced a theory of 'pangenesis', by which individual parts of the parent produced 'gemmules' which travelled to the reproductive organs where they were bolted together. A hand's cells passed on instructions on how to form a hand, the stomach's cells passed on instructions on how to make a stomach and so on. We now know that each and every cell of the hundred trillion cells in the human body carries instructions for making an entire human, information held in the form of DNA. Darwin's model seems ludicrous today, but one can see how our model might itself have seemed ludicrous. Having one hundred trillion copies can seem wasteful, a model stuffed with redundancy.

In the 1880s the German biologist August Weismann (1834–1914) discovered the 'germ plasm', the physical material responsible for transferring inherited traits from one generation to the next. The plasm and the 'determinants' (genes) it contained were provided by the parent organism. They were protected from the 'somatic cells' which carried out all other functions, meaning that it was impossible for any change in the 'determinants' to occur in response to the organism's environment. We now use the word 'phenotype' to refer to the traits exhibited by the organism, and 'genotype' to refer to the genetic information within its cells.

The 'Weismann barrier' made it impossible to accept the theory of Lamarck that parents could pass on 'acquired characteristics', that is, traits

---

[7]Adrian Desmond and James Moore, *Darwin* (London: Michael Joseph, 1991), p. 485.

they had gained or lost during their lifetime. A bitch that had had its tail cut off would not produce puppies without tails. The phenotype did not influence the genotype. You either had the genes or you did not. Elaborating this theory around 1900, biologists rediscovered the work of a Silesian monk, August Mendel (1822–84), on variation in pea plants. Mendel explained gene expression, the process which determines whether the information regarding a particular hereditary trait is 'expressed' (causing that trait to form in the offspring) or not.

Mendel's work demonstrated that it was possible to carry a gene, but not display the condition or trait controlled by that gene. Previous models of heredity had assumed a 'blending' approach: fertilize a red-flowered pea plant with pollen from a white-flowered pea plant, so the reasoning went, and the result would be peas producing pink-flowered pea plants. Blending represented a major problem for the theory of natural selection, as it made it hard to explain how mutations could be passed down the generations, causing a slow process of separation from the original species. It seemed more likely that any useful mutation would be diluted and lost through 'blending' with the main, non-mutant population, which far outnumbered any members with the mutation in question.

Mendel showed that red flowered-ness was passed on by a dominant gene and white flowered-ness by a recessive one. In reproduction the offspring receive one gene from each parent. As a dominant gene, any plant inheriting a red gene would produce red flowers, even if it also carried a white gene. As a recessive gene, there would have to be two white genes for the plant to grow white flowers. This explained how two red-flowered parents could end up producing a white-flowered plant, provided each carried a copy of the (unexpressed) white gene. Instead of blending inheritance, it was 'particulate': made up of bits of genetic information. These bits were like cards taken from both parent's decks, cards that had been shuffled and a hand dealt to their offspring.

Genes explained the workings of heredity in a way which made speciation by natural selection of beneficial mutations easier to imagine. Mendel had published his work in the house journal of a local natural history society in Brünn, Moravia, in 1865. Few took note in the wider world at the time, and Darwin never got to hear of it. Mendel's work was recovered at the turn of the century, but it took until the 1930s and 1940s for Mendelianism and Darwinian selection to be reconciled, in the form of a so-called Modern Synthesis, itself the title of a 1942 book by Julian Huxley, Thomas Henry Huxley's grandson. For several decades they were seen as rival models. Darwinian selection and newer theories of 'heredity' (later called 'genetics') seemed to be diametrically opposed approaches, the former looking forward, the latter looking back. Under the influence of 'heredity' late Victorian Britons came to see evolution as teaching the importance of protecting the nation or race's 'germ plasm' (a shared body of genes, or genome) from the 'taint' of dysgenic or bad genes.

Playing Huxley's game

FIGURE 2 *Maull and Polyblank,* Thomas Henry Huxley, *1857. This photogravure captures the piercing dark eyes which would later form part of Huxley's mesmerizing lecture technique at the Royal Institution and other venues.*

© The National Portrait Gallery, London.

At 9 p.m. on Saturday, 15 September 1858 the biologist Thomas Henry Huxley's 3-year-old son Noel died. The illness had come on suddenly, just 72 hours before. Huxley's wife, Henrietta ('Nettie'), became hysterical, then numb. The death and the Church of England burial service held at Marylebone Cemetery in Finchley shocked Huxley, angered by the words of the Anglican burial service, which included the lines of St Paul 'If the dead rise not again let us eat and drink for tomorrow we die' (1 Corinthians 15:32). With his closest friends away Huxley vented his anger at the Anglican priest and author Charles Kingsley, an acquaintance who had written him a letter of condolence on his loss. The result was a remarkable exchange of letters which saw the pair discuss the immortality of the soul, the laws of thermodynamics, apes and much else besides. At its heart it was an exchange about authority and limits of knowledge, in a word, metaphysics. But it was also deeply personal. 'I have spoken more openly and distinctly to you than I ever have to any human being except my wife,' Huxley wrote in one of his letters to Kingsley.

Huxley went on to explain in the same letter that he found the words from the burial service offensive because they asserted that only those who believed in an afterlife could feel grief. 'Why the very apes know better,' he wrote, 'and if you shoot their young – the poor brutes grieve their grief out and do not immediately seek distraction in a gorge.' Looking back over his career, Huxley denied that he had been motivated by belief in an immortal soul or hopes of a reward in heaven. He would not lie, would not 'shame my boy'. 'Science and her methods gave me a resting place independent of authority and tradition.'[8] This statement seems to epitomize that view of science as a separate realm noted earlier. Yet Huxley went on to explain that he did not see science as based on 'hard' concepts, that he saw no benefit or interest in going back to *a priori* concepts (concepts based on theoretical deduction, rather than empirical observation), in the attempt to explore what lay behind those concepts.

As a practice, he explained, science required the scientist to accept certain concepts unquestioningly, without troubling himself. Those concepts might seem fantastic or unbelievable when inspected closely. The doctrine of the immortality of the soul, he noted, 'is not half so wonderful as the conservation of force – and the indestructability of matter. . . . But the longer I live, the more obvious it is to me that the most sacred act of a man's life is to say and to feel "I believe such and such to be true" '. Huxley's comments certainly chime with Darwin's view that the concept of species was central to his work, even if one could not define what a 'species' was (biologists still cannot).

Huxley used a simile, comparing his scientific practice to a game of cards:

> This universe is, I conceive, like to a great game being played out and we poor mortals are allowed to take a hand. By great good fortune the wiser among us have made out some few of the rules of the game, as at present played. We call these "laws of nature" and honour them because

[8]Huxley to Kingsley, 22 September 1860. Imperial College, London. Huxley Papers, Gen. Letters IX, f. 169.

> we find that if we obey them we win something for our pains. The cards are our theories and hypotheses, the tricks our experimental verifications. But what sane man would endeavour to solve this problem: given the rules of a game and the winnings to find whether the cards are made of pasteboard or gold leaf?

Huxley went on to acknowledge that there were several different metaphysical models. Some held that there was only one 'x' or 'substratum'. Everything in the universe was matter, or everything was spirit. The former belief was materialism, the latter pantheism: God was nowhere, or everywhere, 'according as you term it heads or tails', he added, cheekily. Other metaphysical systems saw the universe as divided into two 'x''s, into matter and spirit. For his part Huxley had made a decision to adopt a materialist model, but he was only a methodological materialist. He was a materialist because it allowed hypotheses to be advanced and tested. He was uninterested in finding out what 'x' was. After all, he concluded in the same letter, 'who knows when the great Banker may sweep away table and cards and all – and set us learning a new game? What will become of all my poor counters then?'[9]

Given his reputation as 'Darwin's bulldog', as a fierce antagonist of churchmen meddling in science, it is surprising to find Huxley adopting such a 'weak' view of the authority of fundamental 'laws of nature' like the conservation of force (the first Law of Thermodynamics). Yet Victorians did not just debate what they thought they knew about the universe's structure and, in the case of evolutionists, the 'laws' they believed guided evolution. They also debated how much *could* be known and with what level of certainty. They even debated the relationship between knowledge (and 'science' comes from the Latin word for 'knowledge') and nescience ('not knowing'). Sometimes, like Huxley, they insisted on a sharp border between the two: either we knew something or we did not. Sometimes they spoke in terms of probabilities: something was more likely to be true than not. Sometimes they conceded that knowledge was dependent on prior belief in certain entities that could never be proved to exist.

In the card game simile Huxley seems to be describing observations ('this finch's beak is thick', 'this finch's beak is long and thin') as cards. As a scientist, as a player of his game, his job is to collect such cards and play them. Huxley refers to his theories or 'hypotheses' as 'tricks', referring to a sequence of cards played in whist, bridge and other card games. In trick-taking games the number of points one earns depends on the order in which one puts down one's hand. Huxley imagines the players beginning to play

[9]Huxley to Kingsley, 22 May 1863. Imperial College Archives, London. Huxley Papers, Gen. Letters IX, f. 229. Reproduced in Leonard Huxley (ed.), *Life and Letters of Thomas Henry Huxley*, 2 vols (New York: Appleton, 1900), 1: 350.

without knowing the rules of the game, without knowing, say, which suit (spades, clubs) are 'trump cards' or even the values of numerical cards. Only the dealer/banker knows the rules.

To 'win', the players have to assume that the dealer is not giving out points randomly, that there is a logical relationship between the cards laid down and the points awarded. As countless psychological experiments have shown, humans are hardwired to relate something that we observe ('effect') to something which happened before ('cause'). We can be tricked to perceive order where there is none. To continue with the Huxley's card game, the players would learn by comparing the winnings they get after playing different cards. Gradually they will learn that the card with a symbol which looks like two circles on top of each other is worth more than the one with a symbol of a single line running vertically. Eventually they will understand the relative values of all the cards and the rules of the game. They will gain in understanding and confidence. They will begin to feel at home in the game.

Huxley's game reminds us that science operates within a model of the universe, not the universe itself. As such, it is dependent on a group of individuals agreeing to follow a set of conventions which themselves constitute 'science' for that group. Given a set of conventions, hypotheses can be shared, tested and discussed. Having testable theories lies at the heart of that set of conventions known as 'scientific naturalism'. Theories of transmutation that rely on a supernatural being (such as Intelligent Design) are not playing this game, though they may be playing another.

Huxley's game also reminds us that any hypothesis is not 'just a theory' – a wild guess – but a key scientific tool. Both Wallace and Darwin were particularly aware that having a hypothesis did not just give them something to do with the observations they made, it actually encouraged them to make more observations than they would otherwise have done. This approach, of formulating a hypothesis, then seeking corroborating facts, went against the method advanced by Francis Bacon in his famous *Novum Organum* (1620), the inductive method most Victorians saw as *the* 'scientific method'. This method presumed a mind clear of any preconceptions whatsoever; a mind which made observations, classified them, and then drew 'inductive generalisations' (general laws or principles) from the patterns identified in the data. Induction began with 'particulars' and made generalizations.

Deduction began by assuming a general principle and then applied it to explain the 'particular' (observations, facts). The more facts it explained, the more powerful and convincing a deduction it was. But deduction could not satisfy that demand for ultimate certainty sought by the Baconian method, it referred to the 'most likely' or 'probable' explanation. Many argued that natural selection should not replace earlier systems unless and until it was proved with ultimate certainty. Yet Darwin noted that mutability could not be proved. In 1861 he praised a review of *The Origin* by the geologist F. W. Hutton precisely because Hutton was 'one of the very few who see that the change of species cannot be directly proved, & that the doctrine must sink or swim according as it groups & explains phenomena'. 'It is really curious how few judge it in this way,' Darwin continued, 'which is clearly the right way.'[10]

When Wallace read *The Origin* he wrote to his brother-in-law Thomas Sims that

> It is the vast *chaos* of facts, which are explicable and fall into beautiful order on the one theory, which are inexplicable and remain a chaos on the other, which I think must ultimately force Darwin's views on any and every reflecting mind. Isolated difficulties and objections are nothing against this vast cumulative argument. The human mind cannot go on for ever accumulating facts which remain unconnected and without any mutual bearing and bound together by no law.[11]

Without scientific hypotheses, we are confronted with an unsettling and unsettled universe, a chaos. With them we can begin to create order in chaos. This order is ultimately imaginary – we still don't know what Huxley's 'x' is – yet nonetheless reassuring for being so.

This book is divided into two parts. The first part considers the intellectual, political and other developments which came together to inspire Darwin's discovery, outlines his key works and surveys the immediate response to *The Origin*. To trace the origins of Darwin's thought we need to step back, before the commencement of Queen Victoria's reign in 1837, into the latter half of the eighteenth century. Chapter 1 considers the thought of Adam Smith, William Paley and Thomas Malthus, all of whom influenced Darwin. The following chapter looks at influences from France where the most sophisticated arguments for and against transmutation were being rehearsed in the 1820s and 1830s. Chapters 3 and 4 are somewhat more narrative in structure, describing the *Beagle* voyage, the writing of *The Origin* and the role of Huxley as 'Darwin's bulldog'.

Part two considers the many institutions which provided a home for evolutionary science in the Victorian period, a period in which the infrastructure of school and university courses, museums and research funding was only just being built. Although it will address such familiar institutions, it also looks at other spaces for evolutionary theorizing and debate, such as the working man's pub, as well as the more distant shores of empire. Here the focus extends beyond Darwin to other great Victorian evolutionists, some of whom we have already met: Kingsley, Huxley,

---

[10] Darwin to Hooker, 23 April 1861. Darwin Correspondence Database, http://www.darwinproject.ac.uk 3098 (accessed 2 May 2013).

[11] Alfred Russel Wallace to Thomas Sims, 15 March 1861. Cited in George Beccaloni, 'Wallace's Annotated Copy of the Darwin–Wallace Paper on Natural Selection' in Beccaloni and Smith (eds), *Natural Selection and Beyond* (Oxford: Oxford University Press, 2008), pp. 91–101 (99).

Wallace, but also Herbert Spencer. All embraced Darwin's ideas, albeit in very different ways, deploying them in a range of contexts: spiritual, philosophical, literary, historical and political.

Kingsley saw *The Origin* as an invitation to create a 'natural theology of the future', bringing an idiosyncratic if nonetheless Christian gospel of evolution to young and old through his fiction and his racist historical writing. As the man who made watchwords of 'evolution' and 'survival of the fittest', Spencer's analogies of development were astonishing in their scope. Yet he has come down to us as the father of social darwinism, the defender of an extreme form of laissez-faire capitalism. Given the relative paucity of scholarship, he remains something of a puzzle. Wallace's socialist-spiritualist model of evolution provides salutary evidence that transmutation could be made to serve left as well as right. Co-discoverer of natural selection, Wallace outlived both Darwin and the nineteenth century itself. His early twentieth-century reflections on our period, 'its successes and failures', afford the perfect place to end our journey.

Further reading

Bowler, Peter, *Evolution: The History of An Idea,* 2nd edn (Berkeley: University of California Press, 1989).

Himmelfarb, Gertrude, *Darwin and the Darwinian Revolution* (London: Norton, 1959).

Hodge, Jonathan and Gregory Radick (eds), *The Cambridge Companion to Darwin* (Cambridge: Cambridge University Press, 2003).

Houghton, Walter E., *The Victorian Frame of Mind, 1830–1870* (New Haven: Yale University Press, 1957).

Ruse, Michael, *Life's Splendid Drama. Evolutionary Biology and the Reconstruction of Life's Ancestry, 1860–1940* (Chicago: University of Chicago Press, 1996).

—, *Darwinism and Its Discontents* (Cambridge: Cambridge University Press, 2006).

Young, Robert M., *Darwin's Metaphor. Nature's Place in Victorian Culture* (Cambridge: Cambridge University Press, 1985).

PART ONE

The Longest Discovery, 1750–1870

CHAPTER ONE

Natural theology

Imagine you were walking across a heath and accidentally kicked a stone. If someone were to ask you how that stone got to be there, you would probably reply that it had been lying there all along. But if you were to find a watch lying on the ground, you probably would not give the same answer to the same question. Whereas the stone appears to be a shapeless lump of rock, the watch is made up of many moving parts, made out of a range of materials: steel, glass, brass, perhaps plastic, quartz and silicon as well. These parts are arranged in such a way that the teeth of one cog turn the teeth of another, so that the electricity flows from the battery and the hands are visible. There is order in this arrangement, as well as in the materials chosen for each individual part. Glass is transparent, but also fragile; steel is strong, but not transparent. If we rearrange the parts or change the materials, this order collapses: the battery shorts, the hands do not turn. If we make the wheels out of glass and the protective face out of steel, the watch ceases to function as it did. The closer we look at the watch, the clearer it becomes that it is the result, not of accident, but of an intelligent mind.

So begins William Paley's 1802 book *Natural Theology, or Evidence of the Existence and Attributes of the Deity, Collected from the Appearances of Nature*. At the time the book was published Paley was Archdeacon of Carlisle. Born in Peterborough to an ordained Anglican priest and headmaster, Paley (1743–1805) had followed his father into the church in his 20s and taught for a number of years at Christ's College, Cambridge, before leaving to marry, raise a family and write books, gradually moving up the ladder of the Church of England. As a teacher Paley saw his role as giving students the questions, rather than the answers. His opening thought experiment is characteristic of this approach. We are presented with a certain object or system and asked 'how did it come about?'

As the introduction noted, it is important to consider the framework within which we pose such questions. For his part, Paley did not adopt the model later championed by Huxley in his letters to Kingsley. He did not see investigation as an activity pursued within a distinct professional sphere, using a set of conventions to create and test hypotheses. Paley certainly did believe that 'cause' and 'effect' existed, rather than being something that human minds imposed on the universe to make it easier to live with. He did not look for demonstrative evidence, that is, he did not see knowledge in terms of either complete certainty or total ignorance.

Following an earlier Anglican scholar-priest, Joseph Butler (1692–1752), Paley recognized that knowledge was limited because human intelligence and understanding were limited. Though a divine intelligence saw certainty wherever it looked, for mortals 'probability is the very guide to life', to quote Butler's 1736 work *The Analogy of Religion*. Thus while physicists puzzle over the nature of gravity, the rest of us continue to walk around without worrying that we are going to float off the ground. The lack of scientific certainty in this, as in many other cases, does not incapacitate us. We don't rope ourselves to the nearest fixed point. It is enough to know from prior experience that we tend to 'stick' to the ground. Our inability to achieve complete understanding does not mean that we abandon any effort to investigate the universe around us, to represent this theory as more likely than that.

Having started with a man-made object, the rest of Paley's *Natural Theology* presents us with a vast array of creatures, organs, instincts, atmospheric phenomena as well as the movements of planets. He demonstrates how each is as complex as the man-made watch, as well as the ways in which each represents a part of one vast system, a watch, as it were, of unimaginable extent and intricacy. How did these individual examples come about? Given his probabilistic approach, quantity is important; each adds to the cumulative force of his overarching proposal:

> Upon the whole; after all the struggles of a reluctant philosophy the necessary resort is to a Deity. The marks of *design* are too strong to be got over. Design must have had a designer. That designer must have been a person. That person is God.[1]

As a discipline, natural theology's main aim was not to prove the existence of a creator God. Even the most advanced thinkers of the seventeenth- and eighteenth-century European Enlightenment retained a belief in a divine Creator. They believed that a Supreme Being had created the world 'in the beginning'. But that did not mean that they believed that this Supreme Being gave human beings (and only human beings) an immortal, immaterial soul

[1]William Paley, *Natural Theology* (Oxford: Oxford University Press, Oxford World Classics, 2006), p. 229.

(destined for a heaven or a hell) or that he had a son named Jesus Christ whom He had sent to earth to redeem humanity from something called 'original sin' (Adam and Eve's eating of the apple in Eden). The Supreme Being might have constructed the universe, wound it up . . . and then left it, like Paley's watch, to wind down in the void. The Being did not necessarily seek an ongoing relationship with humankind. Stories of God speaking to this or that *Homo sapiens* by miraculous means were just that, stories. Though churches had abused human gullibility by encouraging such 'superstitions' for centuries, in the new Age of Reason individuals would come to think for themselves, trusting in the witness of their own senses and intellect, rather than in revelation.

'Revealed' knowledge was supernatural in that it came from outside the natural world, breaking into it. Rather than being the result of human deduction (a kind universe = a kind Creator), this knowledge came in a flash, overwhelming normal thought through a vision, a dream or a voice from heaven heard by this or that prophet or holy man and then shared with others. Chapter 34 of the Old Testament book of Exodus, for example, describes how God appeared to Moses on the summit of Mount Sinai and proposed a covenant (agreement) by which the Jews would receive certain lands in return for obeying a set of rules. These rules – the Ten Commandments – were carved on tablets of stone. Together with the many other revelations scattered across the Old and New Testaments, this revealed knowledge was interpreted as divine guidance on how each individual could live in an ongoing relationship with their Creator.

Natural theology was another means of finding out what God was like; not by hearing His voice, but by making deductions from the world He had created. Was He benevolent or cruel? Did He want his creatures to survive and breed like unfeeling weeds, or did He want living things to experience pleasure in existence for its own sake? Was *Homo sapiens* just one creature among many, or had God taken particular care to provide for this species? The fact that we could find and interpret messages God had left for us was itself a sign of our special status, of the care God had taken for us in providing us with a pleasant place to live.

Some of Paley's examples, it must be conceded, are daft, and his confidence that 'it is a happy world after all' can seem naive. One is occasionally reminded of Dr Pangloss, the fictional professor of 'metaphysico-theologico-cosmolo-codology' in Voltaire's 1759 fable *Candide*. Pangloss certainly believes 'that there is no effect without cause' in this 'best of all possible worlds'. 'Everything having been made for a purpose,' he reasons, 'everything is necessarily for the best purpose . . . Legs are evidently devised to be clad in breeches, and breeches we have.'[2] A keen fisherman, one might have thought Paley would have known not to present life under water as blissful. He notes,

---

[2]Voltaire, *Candide and Other Stories*, trans. Roger Pearson (Oxford: Oxford World's Classics, Oxford University Press, 1990), p. 2.

for example, how shoals of fish constantly rush about and jump up out of the water. 'These are so happy,' Paley concludes, 'that they know not what to do with themselves.'[3] Are they happy, we may well ask? Are their 'frolics' the product of their joy, or constant fear of being eaten by bigger fish? Like Pangloss' 'best of all possible worlds,' so Paley's universe is as benevolent as it is interconnected: the orbits of the planets not only serve to keep the solar system stable, but also provide regular periods of darkness in order to give living things time to rest. 'If this account be true,' Paley notes, 'it connects the meanest individual with the universe itself; a chicken roosting upon its perch, with the spheres revolving in the firmament.'[4]

By the end of *Candide* all have learnt that the world is not without confusion, pain and suffering. Rather than re-evaluate Pangloss, however, Voltaire simply proposes that we find a quiet retreat from this world, a place where we can be still and 'cultivate our garden'. Why bother with Paley, we might therefore ask? One might propose doing so simply on account of the popularity of Paley in early nineteenth-century England. The first 2000 copies of *Natural Theology* sold out almost immediately; there were 24 editions by 1822. One might mention Paley's influence on Darwin. Darwin read Paley's *Principles of Moral and Political Philosophy* (1785) at Cambridge where it was required reading. Darwin attended Paley's old college, Christ's, and even lived in what had been Paley's rooms. Darwin went on to read *Natural Theology*, which inspired his first speculations on the natural world.

But there are other, weightier reasons for paying close attention to Paley's work, however. As the qualifications ('upon the whole', 'if this account be true') indicate, Paley is not working out some neat Panglossian theorem. Far from dismissing alternative explanations for design out of hand, *Natural Theology* gives them a hearing. The book reminds us that a variety of evolutionary theories circulated long before Darwin. Though a creationist, Paley was certainly not out of touch with the latest science. He responds to the latest speculations of English and French men of science. Several of the questions Paley asks in response to such theories merit asking today, which explains why he remains such an important figure in evolutionary thought. Given the fearful climate created by the French Revolution of 1789, it is impressive that Paley was prepared to go so far. The theories he discussed were the same theories many associated with the chaos and destruction that ravaged Europe from 1793 right up to 1815.

Revolutionary appetencies

Among the evolutionary theories Paley considered was that which proposed that 'every organized body which we see, are only so many out of the

---

[3]Paley, *Natural Theology*, p. 238.
[4]Ibid., p. 158.

possible varieties and combinations of being, which the lapse of infinite ages has brought into existence'. The rest had been weeded out 'being by the defect of their constitution incapable of preservation, or of continuance by generation'.[5] This may sound eerily similar to Darwin's theory, but differs in leaving no room for transmutation. Although they emerge over an immense period of time, there is no sense of these species sharing an evolutionary family tree. Paley refuted this theory by pointing to the neatness of the taxonomic order, seduced by the elegance of the Linnaean system. He also argued that such a system relied on spontaneous generation rather than on a single act of creation. George-Louis Leclerc, Comte de Buffon has argued for spontaneous generation of microscopic life forms in works such as his *Histoire Naturelle* (1749), proposing that these organisms then developed into plants and animals, along the lines dictated by their respective 'internal moulds'. Anticipating later Darwinian sceptics, Paley insists that in order to credit this hypothesis we must be able to observe such generation as well as transmutation occurring in real time. He found it difficult to imagine one without the other.

Paley also considered theories on the basis of ideas that life might have emerged as a result of 'elective affinities' or 'appetencies'. In chemistry, for example, certain elements form compounds easily, while others do not. As the adjective 'elective' indicates, they seem to choose which other elements they bond with. We see such an affinity in the nucleotides of our DNA, for example: our DNA can be replicated from a single strand only because of the 'affinity' by which Adenine pairs with Guanine, but not with Cytosine or Thymine. Eighteenth-century thinkers came up with different terms to describe the seeds or germs which might have coalesced from the primordial soup, but many would have shared Paley's view that such theories were so speculative that they were useless. Charles Darwin steered clear of attempting to explain the origin of life, and even today the origin of life receives a fraction of the attention devoted to exploring the origin of the universe as a whole.

Although he takes his time, Paley eventually gets around to the theories advanced by Erasmus Darwin (1731–1802), Charles Darwin's grandfather. Erasmus Darwin was a middle-class Lichfield doctor who had wide-ranging interests in chemistry, botany, meteorology and mechanics. A passionate inventor, Darwin belonged to the Lunar Society of Birmingham, an informal monthly gathering (they met when the moon was full, hence the name) of around a dozen of the leading minds of the Industrial Revolution. Darwin's fellow Lunar men included the steam engineer Matthew Boulton, the manufacturer Josiah Wedgwood and the radical philosopher and chemist Joseph Priestley, the discoverer of oxygen. Although Erasmus had studied at Cambridge, otherwise the Lunar men tended to be self-educated; intellectually and socially detached from the Oxbridge elite, with bright

[5]Ibid., pp. 38–9.

hopes of mankind's potential for improvement, hopes nurtured by French Enlightenment thinkers like Condorcet. Erasmus shared their drive to end slavery as well as the Test Acts, which prevented non-Anglicans from attending Oxford and Cambridge.

In addition to translating Linnaeus into English Erasmus Darwin sought to diffuse a knowledge of the Linnaean taxonomic system among his fellow Englishmen. To render it palatable he depicted the sexual lives of plants in verse, anthropomorphizing their couplings enough to make his *Loves of the Plants* (1789) far more salacious than its title might otherwise have suggested. His *Zoonomia* (1794–6) proposed 'that all warm-blooded animals have arisen from one living filament, which the great first cause endued with animality'. He proposed a model whereby this filament's physiological qualities (which he understood as irritation, sensation, volition and association) would specialize over time, changes being inherited in such a way as to construct an evolutionary pedigree linking the simplest forms of life to humans. Although the grand scheme was only unveiled in a posthumous work which post-dates Paley, *The Temple of Nature* (1803), Darwin's theories were already notorious from his previous works, at least, notorious among the educated elite.

This notoriety derived in great measure from the panicky climate of the 1790s. In the first few years of the French Revolution, which began when a mob stormed the royal fortress of the Bastille in Paris in July 1789, many British observers cheered. It seemed that the French were finally catching up with Britain, securing the liberties and the balanced constitution that Britons had gained in the Glorious Revolution of 1688. Although the speed was dizzying, otherwise the election of a national assembly, the Declaration of the Rights of Man (both 1789) and King Louis XVI's apparent willingness to cooperate in abolishing vestiges of absolutism were uplifting to behold. With Louis XVI's unsuccessful attempt to flee Paris (June 1791), the rise of the radical Jacobin party and the execution of Louis XVI (January 1793), however, the liberal vision retreated.

After the bloodbath of the Terror (1793–4) came a conservative reaction which saw a republic transform itself into an empire. In 1804 a Corsican artillery officer, Napoleon Bonaparte, was crowned emperor in the cathedral of Notre Dame. By that point France had been fighting a largely defensive war for over 10 years (with a short break in 1802–3). A military genius, Napoleon led French imperial forces across Europe, defeating one great monarchy after another until his ill-fated Russian adventure (1812). This marked the turning point. A ragbag collection of armies now pushed French imperial forces back to Paris, restored the French monarchy and exiled Napoleon to the island of Elba. In March 1815 Napoleon escaped and re-formed an army, which was defeated at the Battle of Waterloo by another motley collection of European armies, led by a Briton, Arthur Wellesley, Duke of Wellington.

This was the backdrop against which Paley's *Natural Theology* as well as Robert Malthus' *Essay on the Principle of Population* (1798) were written and read. These years saw Britain at war across the globe, and facing the real danger of invasion, an unfamiliar situation that would not occur again until World War II. French revolutionaries not just proclaimed unsettling ideas of equality, nationalizing the property of the noble elite, but also abolished the Christian church and attempted to set up their own cult of a Supreme Being. Their leaders were well versed in the ideas of French philosophers such as Jean-Jacques Rousseau. They believed that they were fighting for a set of universal ideas and for a vision of a perfect society, rather than for the glory of a single nation. Their armies brought freedom.

It was easy, therefore, for many Britons to suspect those who admired the same thinkers (as the Lunar Men did) – those who believed in human progress of any kind – of being 'enemies within our midst', seduced by French ideas which were all the more menacing for being little understood. Disappointed at how few of his parishioners attended his Ash Wednesday service in 1800, parish priest William Holland was quick to find an explanation. 'Lukewarm, lukewarm,' he wrote in his diary, 'Religion declines yet I trust it will revive again as French Principles begin to be exploded.' Whether anyone in his tiny village of Over Stowey, Somerset had heard of Rousseau (let alone read him) is highly doubtful.[6] But the fears were real, and in some cases the Establishment went on the offensive.

It is hard to believe that mobs of 'Church and King' vigilantes such as that which trashed Priestley's Birmingham home in 1791 acted without support from local elites. In this feverish climate, speculations like Erasmus Darwin's were received as a challenge to order and decency, more of a challenge, perhaps, than his grandson Charles' notions would be in 1859, when Britain's global pre-eminence went unquestioned. For Erasmus' contemporaries 1789 and its aftermath supposedly held a lesson. Speculation on the origin and structure of natural systems could not be safely contained within, say, the Lunar Society or a fashionable Parisian *salon*. They would infect whole societies with false hopes of human perfectibility and destroy belief in the institutions of church, class and legal authority. When the 'utopian' visions collapsed the disappointed mob would, in the absence of any traditional restraints, descend into a state of savagery.

There are points in *Natural Theology*, therefore where one senses that Paley has political as well as intellectual reasons for not considering all of the various eighteenth-century theories of evolution as carefully as he might have. This explains, for example, why he devotes little time to theories which held that 'unconscious energies' represented some kind of life force driving the machinery of life. In a famous 1771 experiment the Italian Luigi

---

[6]Jack Ayres (ed.), *Paupers and Pig Killers: The Diary of William Holland, 1799–1818* (Stroud: Sutton, 1984), p. 27.

Galvani noted that the legs of a frog twitched when an electric current was passed through them. This, together with Franz Anton Mesmer's work on 'animal magnetism', seemed to support vitalism, an ancient theory which argued that living things were endowed with some 'vital force', a non-physical element. Though Paley was dismissive, as we shall see, 'force'-based evolutionary theories came to prominence in the 1870s, thanks to the work of the physicist John Tyndall, and they were also invoked by spiritualist natural historians like Alfred Russel Wallace.

Paley understood that creation was a process of 'drawing out a creation', a process by which God first sets a set of basic principles or parameters and then works within their confines, in a process akin to problem solving. This process was responsible for that vast set of interlinked 'contrivances' (adaptations, we might call them) displayed by this or that specific creature. Had Paley known of the finches of Galapagos, he might well have noted the ways in which each variety of beak was specifically shaped or 'contrived' to break open seeds, catch small insects and so on. Such contrivance, he noted, 'is the refuge of imperfection'. After all, why would an all-powerful God choose to encase nourishing seeds in a tough shell, making such a 'contrived' beak necessary? A cruel God might enjoy making the finch work, but God was benevolent. He had provided a can, as it were, but He had also provided a can opener.

Paley concluded that God had deliberately made things that much harder for Himself in order to display His own intelligence, power and benevolence. Whom was He displaying this to, or for? Humanity alone, as a means of deepening humanity's understanding of a Creator who wished an ongoing relationship. For God this had required establishing a set of laws which governed how the created universe worked. Here Paley was aware of the risk that behind this familiar image of God as lawgiver lay a danger. God, that supreme 'First Cause', could be seen as having delegated His powers to these laws, to a set of 'secondary laws' which enforced themselves without requiring any action or even supervision on His part. Like a watchmaker, God could have organized the various cogs of the universe in obedience to such laws, set the machine going and then departed. Even worse, perhaps, He could be seen as having set limits to His own free will, by creating a machine so perfectly interdependent that to touch any spring or wheel might cause the universe to fall apart. If God was absent or powerless to intervene, what was the point of praying for Him to intervene in His world, to cure this individual's sickness or to end that war?

Was it helpful to refer to the universe as governed by such 'laws'? Was a 'law' simply a relationship between a cause and an effect, or could it be a cause in its own right? This was a debate that would continue through the Victorian period. In his exchange with Huxley Kingsley argued that 'laws' were mere 'customs of matter'. As we shall see, Kingsley was attempting to construct a 'Natural Theology of the Future' with evolution at its heart. Hard-edged laws had no place in such a system, he believed. For his part

Paley did not believe that law could cause anything. 'A law presupposes an agent', he wrote, precisely because it set the parameters within which an agent could act. While he was able to see 'several ranks of agents' operating in the world, God was present in all of them.[7]

Malthus and population

In 1798 another Anglican priest, Robert Malthus (1766–1834), published his *Essay on the Principle of Population*, a book which shared Paley's scepticism of French Enlightenment thought, taking care to rebut the utopian theories recently advanced by William Godwin in his 1793 *Enquiry Concerning Political Justice*. Yet it presented a far more pessimistic view of human superfecundity than Paley had. Paley's *Natural Theology* addressed the created world as a whole, rather than focusing, as Malthus did, on humans. Paley noted the astonishing fecundity of plants and insects, and recognized the apparent role played by chance in determining which individuals of a species survived, rather than being eaten or otherwise destroyed. But he silenced these questions by seeing the quantity of life as related to happiness. The more, the merrier – not only for the swarming fish leaping for sheer joy, but also for us humans, who were thus able to observe God's creative generosity in all living things.

Malthus' 'principle of population' was the opposite of 'the more, the merrier'. Superfecundity meant that human population rose in a geometrical ratio (1, 2, 4, 8, 16), while food supply increased only in an arithmetical ratio (1, 2, 3, 4, 5). Mass starvation, disease and misery would inevitably occur when the rising curve of population crossed the steady line of agricultural output. Though painful, these 'positive checks' were nonetheless part of a divine order which we could not and should not meddle with. Were we to close this or that 'drain' on population, others would open. According to the 'laws of nature', no human had 'a right to subsistence when his labour will not fairly purchase it. . . . At nature's mighty feast there is no vacant cover [i.e. place setting] for him. She tells him to be gone, and will quickly execute her own orders, if he do not work upon the compassion of some of her guests.'[8]

Happily, there were 'preventative checks' on population growth by which we could ensure that we never reached crisis point, the point at which 'positive checks' came into force. These checks were not only prudent, but also moral. Indeed, that 'moral restraint' by which responsible young adults deferred marriage (and hence childbearing) until they were in a financial position to support children was part of Providence, that is, God's plan

---

[7]Paley, *Natural Theology*, p. 24.

[8]T. R. Malthus, *An Essay on the Principle of Population*, ed. Donald Winch (Cambridge Texts in the History of Political Thought, Cambridge: Cambridge University Press, 1992), pp. 248–9.

for humankind. Although eighteenth-century methods of contraception were expensive and unreliable by modern standards, they did exist, and Malthus largely ducked the question as to why they should not be used as a way out of the population trap. He argued that 'improper arts' of contraception would 'weaken the best affections of the heart, and in a very marked manner . . . degrade the female character'.[9] At a period when *coitus interruptus* remained the most common method of contraception, Malthus' equation of premarital sex with childbearing was entirely reasonable.

For centuries the poor had benefitted from alms (charitable donations) given by wealthy individuals concerned about their own souls. Often called 'hospitals', almshouses were institutions specifically structured around the offering of prayers by the poor for a benefactor's soul, it being believed that the prayers of the poor were particularly effective at persuading the saints and Jesus Christ to lift a rich sinner's soul out of Purgatory (the zone in which souls awaited final judgement) and prevent him or her from spending an eternity in Hell. As the name indicated, poor relief was about relieving the worst effects of poverty, not eradicating it. It was certainly not about creating a 'fairer society'. Many chose to interpret Jesus' statement 'ye have the poor with you always' (Mark 14:7) as proof that poverty would always exist. Indeed, poverty could even be viewed in a positive light. Without inequality, noted the eighteenth-century moral philosopher Lord Kames, Christians would not have the opportunity to demonstrate Christian charity.

For Malthus charity and poor relief represented far greater obstacles to 'moral restraint' than 'improper arts' of contraception. They encouraged irresponsible behaviour such as early marriage and the bearing of children whose parents could not support them. They penalized the prudent poor who worked hard and saved for a rainy day. By providing a means of support for single mothers, it encouraged fathers to abandon wife and children: were such artificial supports to be removed 'I scarcely believe that there are ten men breathing so atrocious as to desert them.'[10] Many of these children subsequently died of disease within a year. 'It may be asserted, without danger of exaggeration,' Malthus wrote, 'that the poor laws have destroyed many more lives than they have preserved.'[11]

The first steps towards parish poor relief depended on voluntary demands by JPs (Justices of the Peace, i.e. magistrates) that their neighbours perform their duty as Christians. By the 1570s, however, it had become clear that the state needed to enforce this duty, imposing penalties on those who did not contribute. The Poor Laws were a series of parliamentary acts that culminated in the 1601 Poor Law, passed in the reign of Queen Elizabeth, which remained on the statute books until the Poor Law Amendment Act of 1834. The Poor Laws established a system by which each parish's

---

[9]Ibid., p. 218.
[10]Ibid., p. 265.
[11]Ibid., p. 106.

JP and Overseers of the Poor collected funds from local ratepayers and then distributed them. The Overseers were unpaid volunteers selected from among the wealthier residents of the parish and were presumed to know those who solicited their aid personally. They naturally faced pressure from their neighbours to keep poor rates low.

It was their job to distinguish between the industrious poor, whose poverty was the result of misfortune, accident or incapacity (including old age), and the work-shy, who feigned illness or refused to work even when it was available. It was also their job to ensure that 'vagabonds' and other undesirable non-residents were moved on, in the general direction of their 'home' parish, wherever that was. Although some parishes had workhouses in which to imprison the work-shy and put them to work, most did not. Behind the often hostile rhetoric aimed at the 'feckless' poor, few parishes were willing to stump up the heavy sums required to build a workhouse or 'house of correction'. Instead relief was 'outdoor relief', given partly in kind (e.g. a winter fuel allowance consisting of coal, peat or logs) and partly in sums of money. The recipients remained living in their homes.

Malthus' ideas were shocking not only because he turned notions of Christian giving on their head, but also because they challenged the belief that human life was precious to God and would be provided for, come what may. Had not God instructed both Adam and (after the flood) Noah to 'Be fruitful, and multiply, and replenish the earth' (Genesis 1:28 and 9:1)? Hadn't God provided food to the Jews when they were starving in the desert (Exodus 16)? Although Malthus continued to insist that *Homo sapiens* were uniquely privileged by their Creator (only humans had been given the intellect and will to exercise 'moral restraint'), otherwise he left them exposed, not only to that 'chance' which Paley had cheerfully skated past, but also to the same unrelenting struggle for food. Malthus described those born to families without the means of subsistence as 'the unhappy persons who in the great lottery of life have drawn a blank'.[12]

The New Poor Law of 1834 is often seen as an example of how Malthusian theories encouraged a rising middle class to replace older, paternalist models of face-to-face, 'outdoor relief' with a hard-nosed (perhaps cruel) model of 'indoor relief' in the 1830s. Advocates of the New Poor Law made Malthus their alibi even as the so-called Agricultural Revolution demonstrated his predictions of mass starvation to be groundless. In presenting agriculture as an anchor holding back development Malthus overlooked recent innovations in crop rotation, manuring and machinery. He also overlooked the ongoing project of genetic manipulation which had seen humans engineering higher-yielding crops for thousands of years. Paley held that such human ingenuity would solve any crisis in food supply. He was right. Malthus was wrong.

Malthus' principle was part of a larger vision of the British economy, of wages, rent and profit, one in which manufacturing was sapping manpower

[12]Ibid., p. 66.

and investment from agriculture in order to pack people into overcrowded industrial areas. Malthus accused manufacturers of exploiting the 'cheaply raised population of the surrounding counties', that is, the large pool of cheap labour maintained by the Poor Laws. Manufacturing's 'unavoidable variations' of boom and bust encouraged 'pauperism': a system dependent on poorly paid, unskilled workers who could be hired and fired quickly as demand for manufactured goods rose and fell, with the costs of maintaining this pool in slack times dumped on migrant labourers' 'home' parishes. Urban manufacturers profited in good times, therefore, while in bad times the burden on country Overseers of the Poor increased. Far from representing progress, 'the general increase of the manufacturing system' was unsustainable (in terms of population) and unjustifiable (morally speaking).[13]

Though he credited Adam Smith (1723–90) with helping inspire his 'principle', Malthus disagreed with Smith on the system of export subsidies and import restrictions on wheat known as the Corn Laws. These were heavily criticized by free traders as well as working-class radicals as a 'tax' which took from the poor (for whom bread was a dietary staple) and gave to the rich landowners who owned the vast majority of grain-producing land. Where Smith saw both an unconscionable assault on 'natural liberty' and economic efficiency, Malthus saw something more important than both: a means of maintaining 'a balance between the agricultural and commercial classes of society'.[14] Where Smith hailed a natural 'progress of opulence' from an 'age of shepherds' through an 'age of farmers' to an 'age of manufactures', Malthus asked where the food was going to come from when every society reached the last 'age'. Food was not just one commodity among countless others.

We should pause before dismissing Malthus and his ideas as 'exploded', therefore. Wallace, Spencer, Kingsley and many other leading evolutionists would come to share his concerns in the face of mechanization, industrialization and urbanization. They struggled hard against the Victorian tendency to equate evolution with ever more gadgets and ever-larger cities. Today global warming and associated desertification are leading some oil-rich countries to buy up vast swathes of fertile land in sub-Saharan Africa. It is going to become harder to scoff at Malthusian connections between industrialization, climate change and food security.

The invisible hand

As we have seen, Malthus' political economy was built on the ideas of Adam Smith, professor of Moral Philosophy at Glasgow University. Smith lectured on natural theology at Glasgow, but unfortunately the texts were among

[13]Ibid., pp. 116–17.
[14]Ibid., p. 166.

those lost when his papers were burnt on his death. Smith's first book, *The Theory of Moral Sentiments* (1759), considered how human beings learnt how to be virtuous, not from following a revealed primer (like the Ten Commandments), but by interacting with fellow humans, by sharing passions with one another in obedience to a natural instinct to sympathize. When we expressed any passion (grief, lust, joy), Smith proposed, we were able to determine how much passion was appropriate to the situation at hand by observing the extent to which those around us were willing to 'go along' with us, that is, by expressing the same passion. The correct degree of passion to exhibit was an abstract concept (we never got it 'just right'). It also varied depending on what particular stage of development we had reached. Although savage societies were more violent than civilized ones, savages suppressed their passions rather than displaying them openly, as civilized people did.

Smith's ethics, therefore, can be seen as a sentimental marketplace in which humans 'offer' this or that degree of passion, and are 'bid' a certain degree of passion by spectators. If I stub my toe and 'offer' a furious amount of anger (insert ten expletives here), I will find few takers ('what's the matter with you!?'). Only if I tone it down will my lower 'offer' and the spectator's 'bid' become close enough for an exchange, for that sympathetic sharing of passions ('oh, you hurt your toe? That's too bad . . .') to occur. Humans are innately social, Smith noted, so that all of us constantly seek such exchanges. Indeed, we can only ever feel those passions which we are capable of sharing with others: friends, strangers, or, if we are alone, an imaginary 'impartial spectator'. Sharing passions, even unpleasant ones like grief, bring both actor and spectator pleasure, giving all of us an 'interest' or stake in each other's fortunes, quite independent of any additional ties of kinship, friendship or profit.

In his second, more famous book, *On the Nature and Causes of the Wealth of Nations* (1776), Smith moved from the marketplace of passions to that of goods and services, from a field we might call psychology to one we would call economics. Here again Smith emphasized God-given instincts, specifically 'the disposition to truck, barter and exchange'.[15] These instincts were natural (Smith preferred to write about 'Nature' than 'God'), but were only found in humans; as Smith noted, nobody ever saw two dogs trade. In any community there would be a tendency for workers to specialize, becoming particularly proficient in this or that trade. This would both improve the quality of all goods and services and lead to technological innovations that would increase production, as shown by Smith's example of the pin manufactory (illustrated on the £20 note). Increased production would create a bigger surplus of goods, goods which could be traded in the market for anything else the producer needed. The

[15]Adam Smith, *An Inquiry into the Nature and Causes of the Wealth of Nations*, eds. R. H. Campbell and A. S. Skinner (Indianapolis: Liberty Fund, 1981), p. 25.

larger the community, the more specialized labour could become. Everyone benefitted from free trade, Smith insisted, by lower prices and improved quality of goods and services. Any given exchange did not occur because one party was feeling charitable towards their customer, but only because both sides saw some advantage in it.

Yet Smith had to admit that trade in Britain and Europe was far from being free. Partly this was down to the glory-seeking of princes, who intervened in the economy to subsidize 'strategic industries' (as with the Corn Laws, which subsidized agriculture) to keep as much gold in their country as possible, or to conquer far-flung colonies (Britain was then fighting to hold on to what became the United States), all in the name of national prestige. But mainly it was down to businessmen themselves, who were always colluding with other businessmen to restrict the number of competitors and keep prices high. For centuries guilds had restricted the number of rivals by insisting on lengthy apprenticeships and by deciding who could or could not legally practice a particular trade in a particular town. More recently businessmen had lobbied government to secure protective tariffs, complaining constantly that their sector of the economy was suffering from 'unfair' competition from abroad. These businessmen also colluded to oppress workers, who also suffered from the alienation caused by mind-numbingly repetitive, factory-style production methods.

An observant man with a dry, Scottish wit, in his gloomier moods Smith recognized that expecting free trade to break out in eighteenth-century Europe and its colonies was about as likely as expecting Utopia to be established there. Whether out of a confused belief that gold equalled 'wealth', a desire to protect native industry or even out of a desire to follow this or that philosophical 'system', the state had many 'reasons' to meddle in the economy, to encourage capital to flow this way, but not that, giving subsidies ('bounties') here, putting up a tariff wall there. This system was called mercantilism. Yet, Smith insisted, if they could only leave well alone (in French, *laissez-faire*) then the outcome would be better than anything the wisest statesman could ever achieve.

This was because the sum of individual decisions was greater than its parts, because the market had an intelligence or mind:

> As every individual, therefore, endeavours as much as he can both to employ his capital in the support of domestick industry, and so to direct that industry that its produce may be of the greatest value; every individual necessarily labours to render the annual revenue of the society as great as he can. He generally, indeed, neither intends to promote the publick interest, nor knows how much he is promoting it. By . . . directing that industry in such a manner as its produce may be of the greatest value, he intends only his own gain, and he is in this, as in many other cases, led by an invisible hand to promote an end which was no part of his intention.

> Nor is it always the worse for the society that it was no part of it. By pursuing his own interest he frequently promotes that of the society more effectually than when he really intends to promote it.[16]

Smith only used the phrase 'the invisible hand' once in *The Wealth of Nations*, once in *The Theory of Moral Sentiments* and once in an unpublished 'History of Astronomy'. He may have intended it as a joke, a sop to throw to those 'men of systems' who wanted some sort of demigod or mechanism to put at the heart of their economic models. Smith certainly lived in an age of systems, when French and English utilitarians proposed to measure exactly how much happiness there was in society by, for example, the 'felicific calculus' of the utilitarian thinker Jeremy Bentham (1748–1832). But Smith was as wary of such 'projectors' as he was of men of business. He would not have shared Malthus' belief that Bentham's 'utility' was the only 'test' of what passions should and should not be 'indulged'. Smith did not believe that 'utility' was 'the surest foundation of all morality'.[17]

As Smith noted in his 'History of Astronomy', natural philosophers' models of the universe tended to become simpler over time, in obedience to a human desire for elegance, eventually explaining everything in terms of a single factor or motive (such as 'utility'). Yet this did not mean the model became any better. Unfortunately this process happened to Smith's own economic thought. The 'invisible hand' was extracted and transformed into a fetish, a shibboleth. This did not happen immediately, however, and in the 1790s many Britons associated Smith with 'French Principles'. When an undergraduate named Robert Malthus borrowed *The Wealth of Nations* from his college library in 1788, he was only the second person to do so. By the 1820s, however, Smith joined Malthus and Ricardo as one of the recognized Founding Fathers of classical political economy.

From being a throwaway metaphor Smith's 'invisible hand' now became something to wave sternly at anyone whose notions (be they safety limits on the weight cargo ships could carry or minimum capital requirements for banks) threatened to harm the profits of established businesses. The market was to be trusted to make things right; any restrictions on its 'hand' were deemed to be more trouble than they were worth. Making Smith into the father of a ruthless, dog-eat-dog model of capitalism was not easy. One had to ignore his sceptical view of capital pursued for his own sake, his suspicion of businessmen and overlook his support of state education, death duties and banking regulation. Yet it was done, somehow. 'By what silent revolution of events, by what unselfconscious transformation of thought,' Beatrice Webb wondered in 1886, had a system of thought intended to end 'class tyranny and the oppression of the Many by the Few [become]

---

[16] Ibid., p. 456.

[17] Malthus, *Principle of Population*, p. 282.

the "Employers Gospel" of the 19th century?'[18] Evolutionary theories of 'the survival of the fittest' only encouraged this caricature of Smith and of his capitalist system. Though regrettable, that is not particularly surprising, as *The Wealth of Nations* was a crucial influence on Charles Darwin's development of natural selection. By showing how an 'intelligent' order could emerge out of thousands of decisions by 'dumb' individual agents Smith had proved that, contrary to what Paley said, one could have intelligent design without a designer. That allocation of capital which resulted from free interaction of 'dumb' agents in Smith's market, was more efficient than anything an all-powerful, supremely wise statesman could achieve by top-down, god-like interventions. By boosting output, specialization made an increase in population possible. In the same way speciation (nature's version of the specialization of labour) allowed life to find so many more 'niches' in which to thrive. Though both models relied on instincts or proclivities that Christians could perceive as 'God-given', for some the move from a model in which God regularly and spectacularly intervened in the universe to one in which He acted by 'secondary laws' made thoughts of God seem peripheral.

Once they got over the initial shock, natural theologians learnt to live with Malthus' principle. For John Bird Sumner, Archbishop of Canterbury, it did not represent a tragic barrier, but a means by which God encouraged mankind to develop his higher qualities. Though superfecundity made life into a competition, it was a competition in which only those who worked hard and acted virtuously would breed successfully. In his *Treatise on the Records of Creation* (1816) Sumner wrote

> The operation of this principle [of population], filling the world with competitors for support, enforces labour and encourages industry, by the advantages it gives to the industrious and laborious at the expense of the indolent and extravagant. The ultimate effect of it is, to foster those arts and improvements which most dignify the character and refine the mind of man;[19]

This competition lay at the heart of Christian political economy, which saw property, inequality and competition as Godly spurs to the development of society from 'rude' (undeveloped) tribes to complex, technologically advanced nation-states like Britain. Though God was still revered as benevolent Paleyite Creator, the emphasis on constant activity marked an important shift from a Georgian worldview of genteel, polite stasis to a recognizably Victorian world which admired hustle and bustle almost for its own sake.

---

[18]Cited in Emma Rothschild, *Economic Sentiments: Adam Smith, Condorcet, and the Enlightenment* (Cambridge, MA: Harvard University Press, 2001), p. 65.

[19]John Bird Sumner, *A Treatise on the Records of the Creation, and on the Moral Attributes of the Creator* (London: J. Hatchard, 1816), p. 172.

PHRENOLOGY AND THE CONSTITUTION OF MAN

Phrenology is based on 'cerebral localization', the assumption that different parts of the brain each have their own distinct function. In the late eighteenth century Franz Joseph Gall (1758–1828) identified 27 organs in the brain: each varied in size depending on how developed it was, and these variations could be measured in living individuals by following the outer contours of the skull. In 1818 a young trainee lawyer named George Combe observed a public demonstration of phrenology and was immediately fascinated. Within a year he had published on phrenology, and in 1822 he began lecturing, determined to bring phrenology and its teachings to as wide an audience as possible.

In 1828 he published *The Constitution of Man*, the work that would make him famous. The book was a phrenological primer, identifying 35 organs made up of 'propensities' (such as 'combativeness'), 'sentiments' ('hope', 'wonder') and 'faculties' (of perception, including faculties like 'causality' which posited relationships between external objects). These qualities or characteristics were in tension with one another. Underdeveloped organs could be exercised and so made to grow. Far from being a simple case of this or that organ being too big or too small, however, Combe's organs formed a dynamic system: the goal was to keep organs in balance with one another. Although 'lower' animal proclivities would never disappear, phrenology was optimistic in proposing that all men could enhance their 'higher' moral and intellectual faculties.

Brought up on the thought of moral philosophers such as Smith and Malthus, Combe sought to present phrenology as a complement to their work, providing hard evidence of humanity's passionate nature as well as a means of bringing those passions into line with utility, just as God intended. Critics of phrenology such as William Hamilton, however, accused Combe's phrenology of necessitarianism, that is, of denying free will. One's destiny lay written in one's 'bumps'. If he adopted necessity, however, Combe did not see it as demeaning humanity, but as enabling man to develop his 'constitution' (the dynamic system of different sentiments and proclivities) so that its virtuous nature would become evident.

Phrenology invited action for social reform, rather than a counsel of inactive despair. Combe initially looked to the state to carry out such regulation, treating criminals for their enlarged organs of 'combativeness', for example, rather than fostering violence by seeking retribution through punishment. Though he recognized Smith and other classical economists for revealing the natural laws governing wealth creation, Combe looked to a future in which human productivity would be harnessed within a cooperative community. Similar ideas would find purchase with Herbert Spencer and, in particular, Alfred Russel Wallace, who identified Combe as one of the greatest influences on him. Phrenology provided a model of how to view mankind as both governed by secondary laws, yet endowed with limitless potential.

Paley, Malthus and Smith sought to understand the workings of the natural world. All three did so with reference to a Creator who designed the universe around humanity, who implanted in every man and woman certain passions and instincts which would bring them together in harmonious, happy families and communities. Smith preferred to refer to 'Nature' than to 'God', yet the same anthropocentric (man-centred) view of the universe is clear in his *Theory of Moral Sentiments*. 'The great, the immense fabric of human society' is, he notes, the goal around which the universe is organized. 'To raise and support [it] seems in this world, if I may say so, to have been the peculiar and darling care of Nature.'[20]

Knowledge did not, however, give one power to intervene in the universe's workings or authority to question its laws. All three thinkers emphasized the limitations of the human mind to comprehend the universe, and strongly criticized those who challenged those limitations (arguing for the 'perfectibility' of humanity, as Condorcet did) or who endeavoured to represent the workings of society in what they argued were irresponsibly reductive and mechanistic terms (as utilitarians such as Jeremy Bentham arguably did). Such 'projectors' or fantasists were dangerous because they might mislead others into questioning things which should not be questioned and which could not be altered. That tranquillity which was felt to be central to the pursuit of happiness would be lost, and nothing gained.

Even before the French Revolution broke out in 1789 the political risks of such speculations were clear to British observers familiar with the chaos of England's seventeenth-century Civil War, during which so-called Levellers as well as various radical Protestant sects roused large swathes of the population to protest and take arms with heady visions of democracy, a system widely felt to be inherently unstable. The American and French Revolutions encouraged radical republicans such as the dissenter Richard Price to become more strident, encouraging others who advocated universal suffrage (votes for all adult men) to form societies of 'Friends of the People' across the country. During the long premiership of William Pitt fears that such societies were fostering treason led to a series of parliamentary acts curtailing the very liberties (such as freedom of association) Britain was supposedly fighting France in order to defend. Suspicions among the elite that radical French ideas were infecting the British working classes verged on the hysterical in the 1790s and early 1800s.

In 1790 William Paley delivered a sermon entitled 'Reasons for Contentment', addressed to 'the Labouring Part of the British Public'. Later issued as a pamphlet, Paley's counter-revolutionary text urged the working man to rest content with his lot, rather than 'to covet the stations or fortunes of the rich' and 'wish to seize them by force, or through the medium of public uproar and confusion'. The rich man's ease, idleness and fine food

[20]Adam Smith, *Theory of Moral Sentiments*, eds. D. D. Raphael and A. L. Macfie (Oxford: Oxford University Press, 1976), p. 86.

were not to be preferred, Paley insisted, to the simple pleasures enjoyed by the poor man: the satisfaction of relaxing after a day's honest toil in the fields, the mental stimulus of labour itself and the hearty appetite which gave a relish to the simplest food. Changes in 'ranks and professions' would not bring happiness. Paley himself 'could not get my livelihood by labour, nor would the labourer find any solace or enjoyment in my studies'. 'If we were to exchange conditions to-morrow, all the effect would be, that we both should be more miserable, and the work of both be worse done.'[21] In human society as in the natural world (of which human society was part) God's creatures were designed for a particular niche, and would only suffer if they somehow moved beyond it. As with any science, therefore, natural theology carried heavy political baggage.

Further reading

Bowler, Peter, 'Malthus, Darwin and the Concept of Struggle,' *Journal of the History of Ideas* 37 (1976): 631–50.

Corsi, Pietro, *Science and Religion: Baden Powell and the Anglican Debate, 1800–1860* (Cambridge: Cambridge University Press, 1988).

Eddy, Matthew D., 'The Science and Rhetoric of Paley's Natural Theology,' *Literature and Theology* 18 (2004): 1–22.

Evensky, Jerry, *Adam Smith's Moral Philosophy: A Historical and Contemporary Perspective* (Cambridge: Cambridge University Press, 2005).

Fara, Patricia, *Erasmus Darwin: Sex, Science and Serendipity* (Oxford: Oxford University Press, 2012).

Fyfe, Aileen, 'Publishing and the Classics: Paley's *Natural Theology* and the Nineteenth-Century Scientific Canon,' *Studies in the History and Philosophy of Science* 33 (2002): 433–55.

Stack, David, *Queen Victoria's Skull. George Combe and the Mid-Victorian Mind* (London: Continuum, 2008).

Turner, Michael (ed.), *Malthus and His Time* (New York: St Martin's Press, 1986).

Uglow, Jenny, *The Lunar Men. The Friends Who Made the Future, 1730–1810* (London: Faber and Faber, 2003).

Waterman, A. M. C., *Revolution, Economics and Religion: Christian Political Economy, 1798–1833* (Cambridge: Cambridge University Press, 1991).

Winch, Donald, *Malthus* (Oxford: Oxford University Press, 1987).

—, 'Darwin Fallen Among Political Economists,' *Proceedings of the American Philosophical Society* 61 (2007): 177–205.

---

[21]Paley, 'Reasons for Contentment,' in Paley (ed.), *The Works of William Paley* (Philadelphia: J. J. Woodward, 1836), pp. 496–9 (499).

CHAPTER TWO

Comparative anatomy

In July 1830 the 33-year-old lawyer and geologist Charles Lyell visited Paris. Son of a Scottish gentry family, Charles had become interested in geology while an undergraduate at Oxford. After graduating he combined legal training in London with work studying the chalk formations of Hampshire. This work had led him to France, where a renowned zoologist and palaeontologist, Georges Cuvier, had been studying the chalk formations of the Paris basin for several years. In Paris Lyell could study the impressive collections of the Muséum national d'Histoire naturelle (National Museum of Natural History) and discuss the controversial theories of the Muséum's professor of Zoology, Lamarck. In contrast to Cuvier's static model of life, Lamarck endowed every creature with the power to adapt to changing environments. Lamarck's followers linked this with visions of progressive change, as well as models of development that linked the simplest forms of life (monads) to those considered the most highly developed, such as humans. For Lyell, however, such visions were far too speculative to provide useful tools for the natural historian.

July 1830 was an exciting time to be in Paris, and not just because of the debates over transformism which had raged earlier that year, in which Cuvier's tidy taxonomy had been challenged by a younger generation of Lamarckians. The Bourbon monarchy, which had been restored (twice) with Wellington's help in 1814 and 1815, was in crisis. Having succeeded to the throne in 1824 as Charles X, the king (younger brother of Louis XVI, who had been guillotined in 1793) had pursued a highly conservative agenda unpopular with liberals and the commercial middle class. Choosing to crack down on such opposition, in July 1830 the king issued a series of decrees imposing censorship and disenfranchising large swathes of the population.

The result was revolutionary upheaval in Paris, with vicious street battles for control of royal palaces and other key buildings that ended with the establishment of the July Monarchy, a new constitutional monarchy headed by Charles' cousin, Louis Philippe of the House of Orléans. Though the French would come to refer to the days of unrest in Paris as *les trois glorieuses* ('the three glorious days'), Lyell's response was unenthusiastic. Viewing a rampaging crowd from his windows he wrote home that it would take countless ages for 'Ourang-Outangs to become men on Lamarckian principles'.[1]

What did Lyell mean by his remark? Britons had been comparing Frenchmen to apes for more than 50 years. In the eighteenth century this had been a light-hearted way of pointing to their love of fashionable novelties, their supposed obsession with aping the latest useless accessory or 'kickshaw'. French monkeys were overactive, playful, occasionally impudent. But nothing worse than that. In the 1790s, however, conservative fears of French Principles made the metaphor far more serious, redolent of the wanton destruction and unprincipled, uncontrolled aggression associated with the 1789 Revolution and the Napoleonic Wars that followed. Lyell's comment was partly a joke intended to reassure anxious relatives in peaceful Hampshire, where such revolutionary hurricanes hardly ever happened.

It also referred to a suspicion that theories of natural science could infect the wider public, encouraging them to seek to change the socio-economic environment in which they lived through radical political activity or violence directed against the established authorities. For Paley, Malthus and Smith, natural theology and political economy encouraged mere mortals to trust in the supreme intelligence lurking behind apparent inequality. They taught patience and contentment with one's lot, however modest. Lamarck seemed to have led thousands of Parisians to take their destiny out of God's and into their own hands. The result, Lyell implied, was chaos. Upward development of French monkeys into French men was deferred, perhaps indefinitely. Indeed, Lyell may be implying that, in striving to move forwards, the French have 'reverted to type', moving down the Great Chain of Being, 'back' to the ape 'rung' which, Lyell implies, is their proper station.

Even in an age in which respectability was carefully maintained by all, Lyell could strike contemporaries (including Darwin) as a bit of a snob. As we shall see, Darwin struggled to understand exactly where Lyell stood. Privately Lyell encouraged the younger man's speculations on transmutation. Yet Lyell hesitated to revise later editions of his renowned multi-volume primer, *The Principles of Geology* (1830–33), along Darwinian lines. A respect for the 'dignity of man' may have lain behind Lyell's ambivalent stance: by making all humans (and the French) improved monkeys, theories of transmutation undermined the privileged position of *Homo sapiens*. Not everyone could be trusted to explore the implications of this hypothesis on their own. Perhaps it was best, for now at least, to keep the masses in a

---

[1]Cited in Adrian Desmond, *The Politics of Evolution: Morphology, Medicine, and Reform in Radical London* (Chicago: University of Chicago Press, 1989), p. 328.

position of relative ignorance, as much for their own sake as for that of those for whom they toiled. Meanwhile the elite could be left to get on with refining the hypotheses, working out their kinks and preparing the masses for the changes ahead. Change was inevitable, but did not have to be violent. It could be managed.

Lyell's position can strike us today as dishonest, weak and confused. Huxley's pose of the lonely hero wrestling with (or hugging?) his chunk of odious truth is more appealing. Lyell's concept is characteristic of the British 'Age of Reform' associated with the Great Reform Act of 1832, an age of transition which Darwin lived through. As a number of historians have observed, this age did not see the end of a British *ancien regime*, did not see the wholesale destruction of a religious, political and social order in favour of a new order of religious indifferentism, liberal democracy and utilitarian meritocracy.[2] Reform was not, as Prime Minister Sir Robert Peel noted in his 1834 election manifesto, an invitation to a 'vortex of continual change', but entailed 'a careful review of institutions, civil and ecclesiastical, undertaken in a friendly temper combining, with the firm maintenance of established rights, the correction of proved abuses and the redress of real grievances'.

This chapter tells how the French Principles of Cuvier, Lamarck and their rivals were developed in France in the years 1790–1830, before turning to consider how those Principles were translated and domesticated in Britain in the 1830s. Paris was the capital of natural science in this period. If they could afford to Britons interested in geology, meteorology, botany, zoology, anatomy and physiology made their pilgrimage to the city. They scrounged volumes of Lamarck's *Histoire naturelle des animaux sans vertèbres* and his *Philosophie Zoologique* off friends or reconstructed his ideas from garbled accounts in dictionaries and encyclopaedias. Some even came to revere Lamarck by reading Lyell's hostile account of his views in volume two of Lyell's *Principles of Geology*.

Lamarck and Cuvier

Jean-Baptiste Pierre Antoine de Monet, Chevalier de la Marck (1744–1829) was born in Picardy, in northern France. He served with distinction in the army until injury forced him to retire. He contemplated a career in medicine before an interest in botany led him to study French plants, eventually securing the patronage of Buffon. Buffon had directed the Jardin du Roi (Royal Botanic Garden) in Paris since 1739, and made it an important centre for research. Commissioned a Royal Botanist in 1781, Lamarck travelled in search of rare plants to add to its collection. He subsequently served as

[2]See Peter Mandler, *Aristocratic Government in the Age of Reform: Whigs and Liberals, 1830–1852* (Oxford: Oxford University Press, 1990).

'garde' (a low grade of curator) of the herbarium in 1788. The revolution broke out a year later.

Georges Cuvier (1769–1832), a Protestant, was born in Montbéliard, then part of the Duchy of Württemberg. Educated partly in Stuttgart, Cuvier initially worked as a private tutor in a noble household. Although such positions demanded deference to one's aristocratic employer, otherwise they afforded young men of talent a means of supporting themselves while enjoying a good deal of free time to read and study, with the possibility of securing a pension for life when one's services were no longer required. Adam Smith resigned his Glasgow University professorship (where his income depended on fees, and hence the number of students attending his lectures) to take the Duke of Buccleuch's son on a tour of France. Cuvier used his free time to study natural history (including fossilized creatures), corresponding with other scholars and attending meetings of a local agricultural society. He arrived in Paris in 1795, delivering lectures on behalf of the professor of comparative anatomy at the Jardin des Plantes. He quickly made a name for himself by lectures and papers on mammoths, giant sloths and other megafauna (large animals), securing the first of what would be a series of secretary- and professorships at the heart of the French establishment.

Established under the pre-1789 *ancien regime*, this network of state institutions intended to promote all the arts and sciences grew impressively throughout our period. Established in 1795, a new Institut de France with 144 salaried posts consolidated and expanded earlier, royal foundations, such as the Collège de France (Francis I 1530), the Académie Française (Louis XIII 1635) and the Académie Royale des Sciences (Louis XIV 1666), while the Jardin des Plantes (Louis XIII 1626) was transformed into a Muséum national d'Histoire naturelle. The Muséum was staffed with 12 salaried chairs (i.e. professorships), many of them in specialisms which had never been distinguished in this way before, anywhere.

Though this network of state patronage was highly centralized and a source of considerable national prestige, it remained largely protected from the purges and proscriptions usually associated with regime change. It retained traces of *ancien regime* venality. Multiple salaried positions could be held in parallel, for example, and holders could have their teaching duties performed by substitutes, rather as Anglican vicars had their pastoral duties performed by curates. With natural history included in the new national secondary school curriculum established in 1795, however, France's commitment to the natural sciences as part of a citizen's education was clear. It was a commitment totally absent in Britain at the time, and for some considerable time afterwards.

For a botanist, Lamarck's appointment to a Muséum chair in insects and worms represented something of a new departure. Lamarck remained Buffon's protege, however, in his overall approach to natural science. In his *Époques de la nature* (1780) Buffon had argued that Linnaean classification represented an obstacle to understanding life on earth, because it overlooked life's historical development over time. Along with Linnaeus Lamarck rejected the 'new chemistry' associated with Enlightened philosophers like Lavoisier and Priestley, with their system of elements bound by 'elective affinities', with which they organized themselves into compounds. For Lamarck there were only four elements, with a proclivity not to bond, but to return to a pure, isolated state. Only life could form compounds, thanks to a certain vital essence, a 'highly rarefied caloric' Lamarck called 'the orgasm'.

As the title of Lamarck's 1802 *Hydrogéologie* indicated, his theories of transmutation were founded on the study of meteorology, on the patterns etched in sea, sand and rock by ocean currents, precipitation, erosion and sedimentation. Climate change served as a push factor in Lamarckian evolution, commonly defined as 'the inheritance of acquired characteristics'. Changes in sea level and other environmental factors encouraged creatures to develop particular muscles, organs and faculties. Recent ornithological work, including an edition of Buffon's work in the field, led Lamarck to note how the morphology of birds reflected their different 'needs' or modes of life. Waterfowl took advantage of the membrane between their claws to swim and to decelerate when landing on water. Perching birds had no need for such a membrane. It was more practical for them to have elongated claws with which to grasp the branch to which they clung. The more they exercised and stretched these morphological features, the more pronounced the features would become. When each reproduced, their offspring would be born with slightly larger membranes/longer claws.

By continued use and inheritance over many generations, webbed and long-toed bird feet would emerge. The same process could even be harnessed to explain the emergence of nervous and digestive systems in the simplest life forms. By repeatedly passing along certain routes across the internal space of an amoeba, say, signals or nutrients would etch channels that would harden over time, turning into a spinal cord and an alimentary canal. Although Lamarck was one of the first to use the word 'biology', ultimately he did not see this field as distinct. Whether it traced rivers in the landscape, 'nervous fluid' in the brain or heat exchange in the atmosphere, the natural sciences of meteorology and geology, physiology and zoology, physics and chemistry were all about 'flow'.

Lamarck's biology was materialist in that he saw the universe as containing nothing but matter: no immaterial God, no Spirit, no soul. Such atheism made it highly dangerous. Although Lamarck did not believe that *adult* life forms could change, believing that tissues and channels became rigid when maturity was reached, few readers grasped that restriction. In apparently giving creatures opportunity or room to develop as their own 'will' or 'agency' directed Lamarck advanced notions which were deeply disturbing. Even without the push of environmental factors, creatures could, it seemed, set out in a new direction. Evolution might have an upwards 'direction' of the kind assumed whenever one spoke of 'higher' and 'lower' forms of life. Lamarck tended to see this 'organic movement' (what we would call evolution) in terms of specialization, the emergence of ever-more intricate and complex structures and systems. But this path might be one life blazed for itself, rather than being preordained by the Creator and enforced by His secondary laws. Jacobin orang-outangs could hold a meeting of their species assembly and vote to become humans by means of the general will.

Were species even necessary? Lamarck recognized, as Darwin did later, that all taxa were ultimately figments of the imagination. A natural history which stuck to species naming and species description was a waste of time. It was more important to understand the general principles that caused diversity than 'to commit to memory the names and synonyms of this innumerable multitude of species'. In his *Recherches sur l'organisation des corps vivans* (1802) Lamarck publicly abandoned his earlier belief in fixity of species. 'Now I am convinced that in nature there are only individuals, obliged to modify their ways of life and habits, and thus their organs, in order to survive in a continuously changing environment.'[3]

It is important to emphasize that Lamarckian evolution is distinct from Darwinian evolution, evolution by natural selection. If one were to ask the proverbial woman on the street to explain how she believed evolution worked, one would probably discover a hazy Lamarckianism at work. Though both models easily accommodate selection of this or that morphological feature by 'survival of the fittest', Darwinian and Lamarckian models differ when it comes to the source of these features. Darwin (or, to be precise, the Neo-Darwinian Consensus) says that our morphology is set at the moment of conception. If I am lucky enough to be born with a genetic mutation that gives me large deltoids, my offspring will inherit them. But if I don't, no amount of exercise in the gym will give my children large deltoids. For Lamarck, by contrast, 'use' of organs or muscles during my lifetime will have some effect on the morphology of my offspring. If I acquire buff deltoids through use, my offspring will inherit them. 'Disuse' has the opposite effect of 'use': if I inherit an organ and fail to use it, it will shrink, until eventually, after many generations, it becomes a vestigial organ.

Lamarck also differed in seeking to explain the origins of life in natural terms, a project Darwin avoided, either because he felt nothing could be clearly demonstrated or because he feared charges of materialism. Lamarck's evolution was not centred around a family tree, did not therefore propose that life emerged from a single trunk, a one-off Creation. His evolution presumed that life was constantly coming into being and setting off on its upwards journey. Life which flashed into being earlier had more time to rise on this evolutionary escalator than life which emerged later. Higher and lower creatures were not, therefore, connected to each other by a shared inheritance. Work by Étienne Serres and other French embryologists

[3]Cited in Pietro Corsi, *The Age of Lamarck: Evolutionary Theories in France, 1790–1830*, trans. Jonathan Mandelbaum (Berkeley: University of California Press, 1988), pp. 124, 149.

supported such theories, by showing how the embryonic development of brains and other organs seemed to recapitulate (repeat in miniature) the development of the species as a whole, passing through stages in which the organ resembled those of 'lower' creatures. Lamarck's ideas thus encouraged a highly materialist yet nonetheless teleological view of development. They encouraged further work into embryology as well as a degree of credulity in evidence of spontaneous generation in the following decades.

Intellectually the 1809 *Philosophie Zoologique* was the climax of Lamarck's work to establish the evolutionary laws which organized life – which *were* life-as-process. As a publication, however, it was hardly noticed at the time, either by other natural philosophers or by the general public, despite the ongoing popularity of Buffon's works, which had been re-issued by Sonnini de Manoncourt in a monster edition of 127 volumes (1798–1808). Lamarck's reputation stood high, albeit on account of his species-describing papers on invertebrates and molluscs. His style was partly to blame, as was his seemingly outdated chemistry. On a personal level Lamarck had a tendency to annoy potential allies, preferring to wallow in a smug self-righteousness rather than stoop to persuading others. Blindness in later life completed this image of the doomed prophet. Those aware of his theories often encountered them at second hand, in reference works such as Julien-Joseph Virey's *Nouveau dictionnaire d'histoire naturelle* (1803–04, 2nd edn, 1816–19).

In the late 1790s Cuvier had collaborated with another Muséum professor, Geoffroy Saint Hilaire, on a number of papers. One proposed that speciation might have occurred by 'degeneration' from a common ancestor. Given Cuvier's subsequent campaign for 'fixity' of species, this is striking. Cuvier may have felt unable to challenge the Buffonian climate at the time, one which also saw widespread speculation on the origins of life and the possibility of spontaneous generation. After his appointment as secretary (1802) and then permanent secretary (1803) of the Institut, however, Cuvier seems to have begun a carefully choreographed campaign to establish a new science of 'comparative anatomy', one which embraced palaeontology and provided a key to geology, too. Cuvier's patronage, his official reports, eulogies of recently deceased gentlemen of science and control over the syllabi of the new Imperial University (1808) gave him the means to promote and encourage those willing to follow his lead.

They also enabled him to engage in what historians of science call 'boundary work', that is, to police the frontiers of his field, determining who and what methods did and did not count as 'proper science'. Cuvier could be candid about the pleasure he found in shutting down attempts to construct grand unified theories of life's origins and primary laws. 'Systems must be swept away . . . I rejoice whenever one of them is destroyed by a well-observed fact,' he stated in 1805. His ability to weather changes of regime and success in gaining preferment and distinctions caused a certain amount of jealousy, and led to accusations of having adopted sinister, Machiavellian methods to sideline or discredit rivals, including the more senior Lamarck. Yet Cuvier's model of natural history would not have attracted so many adherents had it simply been based on the abuse of institutional patronage. It provided a workmanlike framework within which to encounter, describe, classify and publish species, as well as to make a career. Yet it managed to combine this with a romantic vision of the comparative anatomist as a heroic figure, whose ordering intellect in a sense created the very creatures it described.

Published in four volumes (two in 1800, the remaining in 1805), Cuvier's *Leçons d'anatomie comparée* (*Lessons in Comparative Anatomy*) were structured around organs: organs of movement, sensory organs, digestive and so on. Following a law of 'the correlation of parts', each and every organ and bone of every creature had a functional relationship to all the others. These relationships were so strong that they created a kind of symmetry that allowed the whole to be reconstructed from any one part. An architectural analogy may help here. In Classical architecture buildings must obey rules governing symmetry (if the west wing has seven bays, the east wing has to have seven as well), proportion (if a doorway is *x* feet across, it must be *y* feet high), situation (its position relative to the sun, to hills, to other buildings) and so on. Anyone familiar from these rules is able to view a ruined classical building (a few columns, the vestiges of one wing) and reconstruct the whole in her imagination.

Cuvier's claim (first made in 1798) that he could reconstruct the skeleton, organs, diet and even instincts of a long extinct creature from a single bone was often repeated – and often demonstrated, in his palaeontological work on mastodons dug up in the United States, mammoths from Siberia and 'pterodactyls' (Cuvier came up with the name) from Bavaria. The claim embodied that mixture of scientific rigour and quasi-creative power noted earlier. The Cuvierite comparative anatomist was a cold, detached professional. Unlike Lamarck, he did not engage in speculation or controversy. But to outsiders, however, his powers seemed supernatural, wondrously magical. Cuvier recognized the importance of presenting growing audiences for handbooks and lectures in the natural sciences with an appealing authority.

For Cuvier, the plan came first. 'It is the genera that must furnish characters, and not character that must determine genera,' he proclaimed. In his 1817 book on *Le Règne Animal* (*The Animal Kingdom*) Cuvier published his replacement for Linnaeus' classificatory system, one made up of four phyla (what Cuvier called *embranchements*), based on four basic plans: Radiata (creatures with radial symmetry, like jellyfish, sea urchins), Mollusca (shelled creatures), Articulata (creatures with segmented bodies, like insects, worms) and Vertebrata (creatures with a backbone, including all mammals, birds, reptiles, amphibians and fish). Since Cuvier's day his 4 animal phyla have exploded into 35. Some have been subdivided into multiple new phyla, others have vanished entirely. Two new 'floors' or taxonomic ranks have been introduced 'above' the phyla (kingdom and domains).

Cuvier did not see any of his phyla as 'higher' or 'lower' than the other. This was an important step in that it broke the 'chain of being', which had seen 'simple' life forms like sponges and jellyfish at the bottom and 'complex' vertebrates at the upper end, with *Homo sapiens* at the apex. Otherwise, however, it ruled out any 'transitional forms' between phyla, because each phylum was based on a fundamentally different unit. As any child knows, it is possible to make boats, airplanes and buildings out of Lego. This diversity is based on a balance between flexibility and absolute precision. The paired button-and-socket connectors (5mm wide, 1.7m deep) of these 'Automatic Binding Bricks' are manufactured with a tolerance of 10 micrometres (a micrometre = 1/1000th of a millimetre) in fairly rigid plastic (called ABS); a finer tolerance or a more rigid plastic (like cellulose acetate, its designers' first choice) would make it difficult to disconnect blocks once they had been connected.

Lego cannot be combined with Meccano, and not just because the latter has its connectors 12.7mm apart, rather than 3mm. Though one can construct boats, airplanes and buildings out of Meccano, the unit of construction consists of plates, not blocks, and bonds are made with screws and bolts. Though the analogy between a Cuvierite phylum and children's toys is imperfect, it does help explain how the former could reconcile simple, refined structures with the obvious diversity of life. It also explains why Cuvier struggled to imagine 'transitional forms'. It was like trying to make something out of a mixture of Lego blocks and Meccano plates. Or, even harder, to imagine something that was part Lego block and part Meccano plate.

Cuvier's work on the Paris basin led him and his collaborator Alexandre Brongniart to see the planet as having passed through a series of violent changes, or 'revolutions'. Whole continents were lifted up, only to be submerged by floods: although the word 'revolution' had gained new, more violent associations after 1789, Cuvier's use of the term nonetheless retained something of the pre-1789 sense: 'revolution' as one of a series of changes which led in a circle. Though these cataclysms wiped out large numbers of species, Cuvier did not pause to reflect on the implications for natural theology. Cuvier did not see his geology as a means of testing or confirming theology, though he conceded that the account of the Flood recounted in Genesis might have been based on one of his floods. In the end these theological or 'philosophical' musings were not part of natural history. A leader of the protestant Consistory who reformed the training of Protestant clergy in Paris, Cuvier's own religious beliefs remain mysterious. The fact that his daughter Clémentine prayed for his conversion suggests he did not have a personal faith.[4]

Cuvier's much-admired and widely translated *Discours préliminaire* (*Preliminary Discourse*) to his 1812 book *Recherches sur les ossemens fossiles*

---

[4]Dorinda Outram, *Georges Cuvier: Vocation, Science and Authority in Post-revolutionary France* (Manchester: Manchester University Press, 1984), pp. 144–5.

de quadrupèdes (*Researches on Fossil Quadrupeds*) challenged Lamarck to provide evidence of transformism, that he point to the 'transitional forms' which must exist if his theories were to be proved correct. Cuvier's challenge would linger on, despite its obvious problems as a test for transmutation. As Darwin would later note, the 'intermediate' form (or 'missing link') between, say a gorilla and a human, was their common ancestor, a long-extinct species that probably didn't look much like either of its descendants, and certainly would not fit any preconceived expectations we might have of what a 'gorilla-man' should look like.

Lamarck adopted a uniformitarian view of the planet's geological history; that is, his model did not call for Cuvier's special effects, emphasizing slow, incremental changes over long periods of time rather than epic cataclysms the like of which had never been recorded by humanity. Lamarck did not see evidence of massive extinctions. The fossil molluscs he had studied were largely still around, though some had changed. When Geoffroy Saint Hilaire returned from Egypt with fossilized cats Lamarck signed the Muséum's 1802 report, which took these 3000-year-old felines as evidence of fixity. Lamarck subsequently changed his position, however, arguing that 3000 years was too short a period in the grand scale (deep time, as we would call it) to serve as evidence disproving transmutation.

Geoffroy Saint Hilaire would eventually prove a stalwart supporter of transmutation, particularly after Lamarck's death in 1829, when Geoffroy tackled Cuvier in a debate on fixity. Geoffroy's *Philosophie Anatomique* (1818–22) argued for 'unity of composition', suggesting that all life was organized historically in a series, rather than being divided up into four distinct divisions, organized functionally, without 'transitional forms', vestigial organs or anything else to clog the white spaces between phyla.

Crossing the channel

Compared to the network of institutions in early nineteenth-century Paris, the natural sciences in London were neglected. A British equivalent to the Académie des Sciences existed in the form of the Royal Society of London (est. 1660), but this body was in a semi-comatose state in 1815. Suspicious that royal patronage might foster a culture of absolutism, Britons had traditionally looked to the less centralized, voluntary model of the Society of Arts (1754, now known as the Royal Society of Arts). This Society offered 'premiums' (prizes) and provided a forum for middle-class and elite amateurs to discuss improvements to agriculture and manufacturing design and technology. Parliamentary support of the arts and sciences was fitful, exhibiting a Georgian parsimony which Victorians would inherit, and celebrate as 'thrift' and 'economy'.

The closest thing London had to the Paris Muséum was the British Museum (BM). Established in 1753, this institution owed its origins not to a magnanimous royal patron but to a private individual, the physician Hans

Sloane, who left his collections of books, natural history and archaeology to the nation. A lottery was held to raise the funds necessary to buy an old seventeenth-century mansion in Bloomsbury in which to house them. Over the next 60 years funding barely sufficed to keep its collections intact, let alone to catalogue and display them in a fashion that might actually advance knowledge. By 1820 this attempt to gather all the sciences under a single roof looked as ludicrous as the famous stuffed giraffes which loomed on its staircase, where they were displayed among unrelated artefacts, for the want of a more fitting home.

Unless there was a clear benefit to trade (as in the case of the discovery of the longitude, which helped shipping), the British regime was uninterested in promoting the natural sciences, preferring to place its faith in the public-spiritedness of amateurs and the enterprise of the free market. This position seemed reasonable, even patriotic, in 1815. It was common to refer to the lexicographer Samuel Johnson's comparison of his famous English dictionary, a much admired two-volume work which first appeared in 1755, with the multi-volume French dictionary prepared by the Académie, which, for all the resources it enjoyed, struggled to get past the letter 'A'. As well as being less 'despotic' and prone to jobbery (corruption), the British funding model (which involved almost no funding at all) was felt to reflect native individuality and common sense, which many taxpayers (with one eye on their pocket) considered preferable to large, expensive institutions pursuing arcane forms of knowledge no gentleman would ever need, and might even regret having, especially if (as was suspected) they were a Trojan Horse for materialists and revolutionaries.

In 1807 a group of gentlemen, several of them fellows of the Royal Society, got together at the Freemasons' Tavern in London to form the Geological Society. From the beginning the society sought to define geology as empirical rather than speculative, safely ensconced within natural theology. The first mammoth skeleton had reached London from America in 1802, and was soon joined by plesiosauruses and other sea-dwelling creatures collected on the Dorset coast by Mary Anning. How did these fossilized 'dragons' fit with scriptural accounts of the natural world? Robert Jameson's popular but heavily edited 1813 translation of Cuvier's *Discours preliminaire* denied that Cuvier's account of the earth's history contradicted scripture. The Rev William Buckland, reader in Geology at Oxford University (1784–1856), entitled his 1823 account of fossils found in Kirkdale Cave, Yorkshire *Reliquiae diluvianae* ('Relics of the Deluge'). Buckland's reconstruction of the habits and habitat of extinct Yorkshire hyaenas was the closest pre-1832 Britain had to offer by way of set-piece to stand comparison with the many anecdotes of Cuvier's magical abilities to summon long-lost creatures from scattered remains.

There was a romantic, sometimes playful element to the way in which Buckland and the author of *Fossils of the South Downs* (1822), the surgeon Gideon Mantell (1790–1852) brought such creatures to life in the 1820s. In Buckland's whimsical lectures, in his and Mantell's published works, as well as in the illustrations of lost worlds prepared by John Martin and Henry de la Beche, earth's prehistory was divided into multiple ages in which top predators fought for predominance, their writhing bodies gleaming in mid-bite, before they were wiped out by another display of divine wrath. Such was the fascination for these natural dramas that by the 1830s the geological threat to scripture seemed beside the point. In any case, did the phrase 'in the beginning' refer to the beginning of all time, earth time or human history? What with such variants and the 'plurality of worlds' thesis, it is impossible to pin all but a few cranks down as 'literalist' or 'non-literalist' in their approach to the relationship between Genesis and geology.

Unable to support himself even on two university readerships (in mineralogy and geology), Buckland had considered leaving teaching before the offer of a lucrative canonry at Christ Church in 1825 changed his mind. Such posts were only open to Anglican priests and were rarely used to support men of science. Oxbridge was not the only world in which patronage controlled appointments, rather than individual merit. The same was true of the medical schools of London and Edinburgh. Renegade professors like Robert Edmond Grant (1793–1874), London University's first professor of Comparative Anatomy, and his close ally, the journalist Thomas Wakley (1795–1862), made Wakley's periodical *The Lancet* a sounding board for criticism of such sclerotic institutions as the Royal College of Physicians, the Royal College of Surgeons (RCS) as well as the BM.

The Lancet spoke up on behalf of the growing number of trainee doctors and surgeons who found themselves excluded on religious or class grounds from the Colleges, who saw the plum jobs at London hospitals filled, not by any meritocratic system, but by individuals 'qualified' simply on account of being born into the same Oxbridge Anglican elite. Having been apprenticed to his uncle at Guy's Hospital, Sir Astley Cooper had become a surgeon and lecturer at Guy's and St Thomas', where he provided salaried posts for no less than seven nephews and godsons, several of whom in turn made it onto RCS' governing Council. The many articulate young men who did not have the good fortune to be related to Sir Astley Cooper either trained in private medical schools in London or travelled to Paris. Though the training they received in such schools was in some respects superior to that available in the 'official' hospital-based schools of London, these students nonetheless found their certificates rejected by the RCS as insufficient preparation. Only those who studied in the 'official' schools could sit the RCS' exams, a legal requirement to practise surgery.

Such pressure would eventually lead to parliamentary select committee investigations into medical education (1834) and the BM (1835), as well as the formation of the reformist British Medical Association (1836). In the meantime Francophilic radicalism seized the student body, especially at the time of the 1830 July Revolution, when tricoloured leaflets did the rounds of lecture rooms, encouraging concerted action against professors felt to be too conservative. One London professor saw his humble demonstrator elevated to adjunct professor. Professor of Physiology at King's College London, Charles Bell (who was also professor of Anatomy at the RCS) resigned in November. A Paleyite who produced an edition of *Natural Theology*, poor Bell never completed his course of lectures on 'design'. His reverence for 'design' and his Scotch accent were lampooned in *The Lancet*. But Bell was knighted the next year.

Having trained and qualified as a physician in Edinburgh, in 1815–17 Grant had travelled to Paris and visited the Muséum, returning at regular intervals thereafter. He incorporated Lamarckian ideas in his London University lectures to students. Grant's inaugural lecture as professor of Zoology described his field of study as investigating 'the origin and duration of entire species, and the causes which operate towards their increase or their gradual extinction . . . and the changes they undergo by the influence of climate, domestication, and other external circumstances'. Cuvier's 'revolutions' were, he stated, contrary to 'science, facts, and human reason'.[5] Grant also spread these theories outside the university, contributing articles to the *Edinburgh Journal* and *Edinburgh New Philosophical Journal* in 1826–27 and delivering lectures aimed at the educated general public, including at the Zoological Society in 1832–33. J. H. Green and the Paris-educated Jones Quain lectured on Lamarck and Cuvier in London in 1824 and 1830; the latter served as professor of General Anatomy at London University in 1831–35. Grant's *Outline of Comparative Anatomy* (1835–41) also trumpeted Lamarck's ideas and was available in cheap paperback. Grant also served, briefly, as one of Darwin's mentors in Edinburgh in 1826. Though Grant lived until 1874, he was written out of the history of discovery written by Huxley and his allies.

REFORMING THE BRITISH MUSEUM

To the historian looking for a set piece struggle between Establishment and transmutationist renegades for control of a state-sponsored scientific institution, the 1835 select committee on the BM is a dream come true. The committee consisted of members of the House of Commons, parliament's lower house, appointed to gather evidence and report back to the House on a specific question. In the Age of Reform such committees came into their own as a means of carrying out Peel's aforementioned 'careful review of institutions, civil and ecclesiastical, undertaken in a friendly temper' (though this 'temper' was often lacking). Committee proceedings were transcribed verbatim, and are now fully digitized, allowing us the sensation of being 'in the room'.[6]

---

[5]Cited in Desmond, *The Politics of Evolution*, pp. 66, 300.

[6]See *House of Commons Political Papers*, http://parlipapers.chadwyck.co.uk/marketing/index.jsp. (accessed 2 May 2013).

The committee on the BM was chaired by a radical, Benjamin Hawes, a soap manufacturer who represented a plebeian London constituency. But the committee also included a High Church Tory MP, Sir Robert Inglis, who viewed the Whig Party's attempts to diffuse what was called 'useful knowledge' among the working classes through cheap educational newspapers and free museum access sceptically. As he observed in a later (1844) Commons debate on free museum admission, there was no 'intimate connection between intellectual and moral goodness'.[7] Expertise was not what the BM required; the leadership provided by its trustees (the Archbishop of Canterbury, various aristocrats and others with limited knowledge of natural history) should be allowed to continue to lead it.

Robert Grant was Hawes' star witness, and the two had almost certainly worked out in advance what questions the latter would ask and what answers the former would give. The former pushed the French model: a new structure of departments with instructors to give public demonstrations, whose heads would serve as paid governors, reporting direct to a ministry, not as amateur trustees. Grant criticized the BM's Assistant Keeper, John Children, for not having a single fossil ammonite on display despite the richness of the institution's collection. He also blamed him for the lack of labels and poor arrangement. The museum's holdings should be arranged to display 'the whole continuous chain of beings, from the lowest corals up to the highest animal forms that exist', along transmutationist lines.

The Tories fought back with, Francis Egerton (later Earl of Ellesmere) questioning Grant's right to be considered an expert in zoology. Comparative anatomists and zoologists from St Thomas' Hospital and the Zoological Society of London argued that the correct arrangement was the Cuvierite arrangement. Robert Owen argued against any overhasty adoption of new theories. Many of these experts owed their jobs to the conservative members of the committee and the boards of the BM, RCS and other learnt societies. Owen had enjoyed stays at Egerton's Oulton Park estate in Cheshire, admiring the aristocratic dilettante's collection of Old Masters and fossil fishes.

In the end Egerton and Inglis won. The only changes made to the BM were the division of the Natural History Department into three (Botany, Zoology and Mineralogy) and minor revisions to salaries and acquisition budgets. In 1856 Owen would be appointed superintendent of this department, whereupon he advocated its relocation to a separate, purpose-built home in South Kensington. One might have expected him to have complained of the BM's many deficiencies in 1834. Unlike Grant, Owen was intent on working within existing patronage networks, rather than trying to assault them from outside or build his own. As it happened, the ranks of Radical MPs were decimated at the 1837 General Election. Owen was playing the long game.

---

[7]16 April 1844. *Hansard* 3rd Series, vol. 64, 38.

The Owenite settlement

Born into a middle-class merchant family, Richard Owen (1804–92) briefly attended Edinburgh University (1824–25) before beginning his apprenticeship at one of the London teaching hospitals, St Bart's. He supplemented his income by assisting the conservator of the RCS' Hunterian collection. This enormous array of anatomical specimens (human bones and preserved organs as well as preserved reptiles, mammals and fish) had been formed by the leading eighteenth-century anatomist John Hunter. On his death in 1793 the RCS had been charged with preserving and cataloguing his collection, which he had munificently donated to the nation. Thanks to Wakley's *Lancet* the RCS' neglect of the Hunterian became something of an embarrassment in the 1820s. In 1830 the Council hired Owen to catalogue the collection, a task which took 26 years, and which involved re-housing the collection in a purpose-built gallery at the RCS' home in Lincoln Inn Fields, London.

Owen did a fantastic job, considering how poorly the RCS paid him, and how little they appreciated what he was endeavouring to achieve, not only for those fields close to their professional interests (such as physiology and pathology), but also for other natural sciences. By 1849 he claimed that the Hunterian Museum was 'the National Depository of the Collections illustrative and forming the groundwork of the Sciences of Physiology, Pathology, and the Comparative anatomy of Animals both existing and extinct'.[8] He might have added palaeontology and geology to the list. Under his management the collection gained a number of specimens and casts of living and extinct megafauna, including the extinct giant sloth (the myolodon). He faced increasing resistance from inside the RCS, however, as well as from *The Lancet*, both of whom wondered how ancient sloths helped improve the practice of surgery on humans.

Although accusations of flunkeyism would hang around Owen right up to his death, *The Lancet*'s opposition in the 1830s was primarily due to Owen's Cuvierite approach to his cataloguing, his emphasis on Paleyite functionalism. By bringing order to the Hunterian (where visitor numbers increased after the rebuild), by publishing papers and delivering public lectures on exciting, charismatic 'new' fossil creatures like the mylodon, moa and iguanodon as well as on the exotic living creatures displayed at the Zoological Society, Owen had pursued a rather Peelite agenda of steady reform of abuses, avoiding the 'vortex' of notorious French theories of transmutation and raising the profile and prestige of British natural history. Indeed, Robert Peel secured Owen a Civil List pension in 1842 and in 1844 commissioned John Linnell to paint Owen's portrait, which Peel hung

[8]Cited in Nikolaas A. Rupke, *Richard Owen: Victorian Naturalist* (New Haven: Yale University Press, 1994), p. 27.

opposite Cuvier's in his picture gallery at his country estate, Drayton Manor. Owen also secured a free grace-and-favour residence from Queen Victoria, Sheen Lodge in Richmond Park.

In the late 1850s Owen would be picked out by Huxley as his arch-nemesis. Huxley did not just make history through his activity as 'Darwin's bulldog', he also wrote the first draft. Ignoring Owen's support early in his own career, Huxley would portray Owen as a stubborn anti-transmutationist ready to die (and tell untruths) in the last ditch for a Paleyite, Tory Oxbridge elite. It is hard not to view Owen's pre-1859 career in this light, but we can try. It is important to note Owen's use of the concept of the 'archetype', of a central, ideal form with a transcendental rather than purely functional significance. Although Owen may have kept such views to himself until the early 1840s, he does seem to have been heavily influenced by what the Germans called *Naturphilosophie*, a metaphysical understanding of the organic unity of the universe that Cuvier himself reviled. Indeed, Owen may have encouraged a tendency to continue thinking in such ways long after the Germans themselves had abandoned *Naturphilosophie*.

Naturphilosophie had originated in late eighteenth-century German Romanticism and numbered the great poet, playwright and polymath Wilhelm von Goethe among its disciplines. In Goethe's *Faust Part One* (1808) he had a character note the tendency to lose the organic essence of observed phenomena in one's very eagerness to take it apart or dissect it. 'He who endeavours to recognize and describe something alive / First seeks to expel the spirit [*Geist*] from it / Then he has its parts in his hands / Yet the spiritual bond [*das geistige Band*] is unfortunately lost.'[9] There were limits to Baconian induction, or rather that method was incapable of observing whatever gave a certain arrangement of atoms, chemicals or organs the quality of life, a consciousness.

Goethe and fellow *Naturphilosoph* Lorenz Oken independently developed the vertebrate theory of the skull, by which the skull was understood to be made up of a number of fused vertebra. Though Cuvier certainly saw a vertebra as a definitive unit of construction for one of his phyla, he scorned these German transcendentalists' attempts to lead all anatomy (plant as well as animal) to a limited number of primeval forms or *Urbilder*, such as Goethe's *Urpflanze* (or 'original plant') or an ur-vertebrate. Relating observable forms of life to such invisible concepts, which were loosely understood as the original 'thoughts' of God, was unscientific. Just as the genera gave the characters, so the function gave the form. For Cuvier.

But not, it seems, for the 'English Cuvier', for Owen. Owen was a correspondent and admirer of Carus, who had written several massive works on comparative anatomy and physiology, who in 1828 had published the

---

[9]Johann Wolfgang von Goethe, *Faust. Der Tragödie erster und zweiter Teil* 1 (Munich: C. H. Beck, 1994). 1936–39.

skeletal plan of the ur-vertebrate. Owen had an even stronger admiration for Oken, whom he met in Freiburg in 1838. Owen organized the translation of Oken's 1831 *Lehrbuch der Naturphilosophie* (*Handbook of Naturphilosophie*) into English in 1847, under the slightly evasive title *Elements of Physiophilosophy*. He had a strong ally in a Swiss-born Harvard professor Louis Agassiz, an expert on fossil fishes and glaciation who, like Owen, embraced both Cuvier and *Naturphilosophie*. 'A species is a thought of the Creator,' Agassiz observed to a student. For Agassiz glaciers acted as God's eraser, maintaining the *cordon sanitaire* between Cuvier's tidy divisions. Different animal species were 'variations of an idea, and not the result of diverse circumstances and influences operating on them'.[10]

THE BRIDGEWATER TREATISES

In 1829 Francis Egerton, the 8th Earl of Bridgewater, died in Paris. An eccentric, obsessive manuscript collector, Egerton had been ordained an Anglican priest, but left pastoral duties to others, preferring to study and publish on the classics and biographical accounts of his titled ancestors. Egerton later moved to Paris, where he gained a certain notoriety as a result of his habit of dressing his dogs for dinner. He would be justly forgotten had his will not allocated £8,000 (around £400,000 today) to commission scholarship on natural theology. The president of the Royal Society was entrusted with the task of appointing suitable scholars, whom he chose with the help of the Archbishops of London and Canterbury.

Between 1833 and 1836 eight titles were published, known collectively as the Bridgewater Treatises, by eight individuals, four of them clergymen. Although the first editions were far too expensive to reach a wide audience, subsequent editions (of which there were many) did achieve the aim of providing curious autodidacts in the United States as well as Britain with 'safe' reading on the natural sciences. The selectors do not appear to have had a systematic division of topics in mind, preferring to find their man first. Among the authors commissioned was the Cambridge don William Whewell, who wrote on *Astronomy and General Physics* (1833) and Buckland (*Geology and Mineralogy* 1836). They also included the hapless Charles Bell, who wrote *The Hand: its Mechanism and Vital Endowments as Evincing Design* (1833). Most had experience of writing for non-specialist audiences, as Bell had with his best-selling *Animal Mechanics* (1827–29).

In Bell's hands Paley and Cuvier merged seamlessly into each other, the latter repeatedly cited and praised for the light his discoveries shed on the Creator's fashioning of the hand into countless 'contrivances' (including swimming

[10]Louis Agassiz, *The Structure of Animal Life* (London: Sampson Low, Son and Marston, 1866), p. 117.

and flying). Although Lamarck was not mentioned, he was nonetheless a lurking presence:

> It is, above all, surprising with what perverse ingenuity men seek to obscure the conception of a Divine Author, an intelligent, designing, and benevolent Being – rather clinging to the greatest absurdities, or interposing the cold and inanimate influence of the mere elements, in a manner to extinguish all feeling of dependance in our minds, and all emotions of gratitude.[11]

The Hand ended with an 'additional chapter' that warned against the dangers Bell saw confronting those who pursued science (as in knowledge) by 'experiment' rather than by 'philosophical enquiry', as 'scholars' rather than as 'philosophers'. The latter's experiments could enfeeble that proper reflex by which those observing the action of secondary laws felt renewed veneration for God. 'Newly acquired knowledge' might serve to 'embarrass, if they do not mislead [the student]; in short, he has not had that intellectual discipline, which should precede and accompany the acquisition of knowledge'.[12]

Though the Bridgewater Treatises gave the reader accessible, well-written and illustrated guides to fields such as geology and anatomy, therefore, knowledge came with stern reminders of how not to abuse it. The lesson was simple: by all means study secondary laws, but don't use them as a bridge to materialism.

Whereas many Britons continued to lump them all together and wish them to the devil, by the 1830s others recognized that when it came to these natural sciences, French Principles did not represent either a unified set of theories or even a unified set of principles by which to advance and test theories. Cuvier's model proposed a descriptive anatomy which rejected any attempt to see this or that branch of his tidy taxonomic tables as 'higher' or 'lower', 'less' or 'more' developed. Though it advanced a natural history, it was a history of clearly defined ages or cycles that didn't lead anywhere. Lamarck saw environmental and morphological change as directly related. He proposed a dynamic anatomy, one of constant transformation of life into 'higher' forms by self-organization.

Had Lamarck and Cuvier been engaged in a debate over fixity? Was Cuvier's rise – or Owen's? – the result of an Establishment plot to quash transformism? It is easier to see them as carrying on a dialogue of the deaf than anything more substantial. As we have seen, each understood natural history to be something very different from the other. Cuvier was thinking strategically about 'natural history' as a set of institutions in a way Lamarck was not. Owen brought a third set of tools from *Naturphilosophie*, including

[11]Charles Bell, *The Hand* (Philadelphia: Carey, Lee and Blanchard, 1833), pp. 112, 114.
[12]Ibid., p. 206.

the imaginary yet compelling concept of the archetype. Rather than adding to a Cuvier/Lamarck conflict, however, Owen helped English comparative anatomy and its associated geology, zoology and palaeontology realize a Peelite middle way between the 'vortex' on the one hand and intransigence on the other.

Although geology and comparative anatomy did not feature strongly in its early deliberations, otherwise the British Association for the Advancement of Science (BAAS) provided an institutional home for this middle way. Founded at York in 1831, the BAAS reflected aforementioned fears that Britain was falling behind continental Europe in the study of the natural world. Like its founding member, the journalist David Brewster, it looked north rather than south to London. Its annual summer meetings changed venue every year and drew extensive newspaper coverage, both of which boosted membership from 353 in 1831 to 2,774 in 1835. Though its leading lights included many Oxbridge-educated Anglicans, including clerics, its spirit was non-sectarian, to the extent of admitting Dissenters (i.e. non-Anglican Protestants). Politically it embraced Whigs (who were slowly becoming Liberals) and Conservatives.

The BAAS' founding generation viewed science as a polite, collaborative activity which encouraged reverence for the Creator as well as patriotic pride. In contrast to the venerable Linnean and other London societies the BAAS admitted women to meetings and, from 1841, to full membership. It distributed small research grants and lobbied parliamentarians. As decades passed it found itself adding to the number of 'Sections' within its big tent, recognizing new specialisms and policing their borders, in a way which arguably came to exclude the curious layman. Even in the early days insufficient levels of mass education and the £1 (£50 today) annual membership would have excluded the vast majority of the British population, preventing the BAAS from being a popular institution. The BAAS nonetheless merited being called 'the parliament of science', reflecting and, in turn, fostering a scientifically literate conversation that was both local and national, and open to men as well as to women of a certain class. It epitomizes what might be called a Peelite vision of science.

Further reading

Corsi, Pietro, *The Age of Lamarck: Evolutionary Theories in France, 1790–1830*, trans. Jonathan Mandelbaum (Berkeley: University of California Press, 1988).

Corsi, Pietro, 'Before Darwin: Transformist Concepts in European Natural History,' *Journal of the History of Biology* 38 (2005): 67–83.

Desmond, Adrian, *The Politics of Evolution. Morphology, Medicine, and Reform in Radical London* (Chicago: University of Chicago Press, 1989).

Gissis, Snait B. and Eva Jablonka (eds), *Transformations of Lamarckism: From Subtle Fluids to Molecular Biology* (Cambridge, MA: MIT Press, 2011).

Hilton, Boyd, 'The Politics of Anatomy and the Anatomy of Politics,' in Stefan Collini, Richard Whatmore and Brian Young (eds), *History, Religion and Culture. British Intellectual History, 1750–1950* (Cambridge: Cambridge University Press, 2000), pp. 179–97.

Jordanova, Ludmilla, *Lamarck* (Oxford: Oxford University Press, 1984).

Morrell, Jack and A. Thackray, *Gentlemen of Science: Early years of the British Association for the Advancement of Science* (Oxford: Clarendon, 1981).

Outram, Dorinda, *Georges Cuvier: Vocation, Science and Authority in Post-Revolutionary France* (Manchester: Manchester University Press, 1984).

Rupke, Nicolaas A., *Richard Owen. Victorian Naturalist* (New Haven: Yale University Press, 1994).

Secord, James A., 'Edinburgh Lamarckians: Robert Jameson and Robert E. Grant,' *Journal of the History of Biology* 24(1) (1991): 1–18.

Sloan, Philip Reid (ed.), *Richard Owen. The Hunterian Lectures in Comparative Anatomy* (London: Natural History Museum Publications, 1992).

CHAPTER THREE

Writing *The Origin*

In 1828 one of the leading doctors of the town of Shrewsbury, Robert Darwin, was busy trying to work out what career was right for his second son Charles, then 19. Charles' grandfather, Erasmus Darwin, had died in 1802. His mother's father, the potter Josiah Wedgwood had died earlier, in 1795. That conservative reaction which began with the 1791 Birmingham riots had continued, and even within the Darwin family this radical, free-thinking heritage had been cloaked in Anglican respectability. Charles had been baptized into the Anglican faith. He attended boarding school in Shrewsbury. Outside school Charles learnt to hunt, shoot and fish.

As the male child of an upper middle-class family, Charles' education aped that of the landowning aristocratic elite. Greek and Latin language, literature and history were important because the classical civilizations of Greece and the Roman Republic and Empire (fifth century BCE to first century CE) represented the pinnacle of human achievement in the arts and in the other useful sciences. Lumped together under the term 'Antiquity', this civilization had seen human society develop from small self-governing city-states like Athens into the vast Roman Empire, itself the result of feats of soldiering and statesmanship that continued to astound. In most cases, however, knowledge of Antiquity was not taught in a very inspiring manner. Charles was not the first British schoolboy to find rote-learning speeches by the Roman emperor Caesar or translating chunks taken from the works of Roman historians like Tacitus boring.

But there was little doubt among parents that such an education taught transferable skills, at least, for privileged boys expected to serve as Members of Parliament, priests, military officers or expected to manage large swathes of farming land. Persuasive rhetoric, manly virtue and other leadership qualities were requisite for all these genteel occupations, which were genteel precisely because they involved governing, commanding, employing or preaching to large numbers of one's social inferiors. Those inferiors had little education before 1870. Though literacy rates were high relative to other European states, otherwise the education they enjoyed ended at 10 or 12 and rarely went beyond a few religious texts and basic mathematics.

The finest orators, patriots and soldiers of all time had been and gone, in Antiquity, and hence it was obvious that they offered the best models to introduce to boys. The expansion of the British empire over the Victorian period made these skills all the more useful. As the headmaster and history professor Thomas Arnold noted in 1842, the torch of civilization had been passed from Greece to Rome. It had then passed to the Germanic tribes who overwhelmed the Roman Empire in the fifth century CE. As the leading living representatives of the Anglo-Saxon race, Britain's destiny as the new Rome was clear to any historian. It was also clear to any natural historian.[1] Any reader of the Bridgewater Treatises knew that God had placed this most favoured race on an island with favourable tides, carefully arranged rivers and plentiful supplies of coal precisely so that Britons could trade, industrialize and take to the seas, building a great empire on a scale to rival the Rome of Augustus. Natural history taught Britons their past, but also their future. Past, present and future were not products of chance, but of Providence: God's plan for the universe.

Although some in the family found him domineering, as head of the Darwin household Robert naturally decided what careers his children followed. He did so without his wife's counsel, as she had died when Charles was 8, and the boy hardly remembered her. Charles was something of a worry. Without decisive action Robert feared Charles would spend his life hunting and fishing – the family was certainly rich enough to support such an idle existence. Charles did not perform well academically at Shrewsbury. Robert sent him to live with his more disciplined brother in Edinburgh, hoping Charles would follow his brother's example and study medicine. Instead Charles was disgusted by dissection, bored by his professors and complained about having to wake up early. In April 1827 he came south again, without a degree and without a plan.

In sending him on to Christ's College, Cambridge University in 1828 Robert signalled that he had abandoned hopes of his second son practising medicine and had decided that Charles was more cut out to be a Church of England priest. Like Oxford, Cambridge University was a medieval institution originally established to train priests, and it was only open to members of the Established Church (i.e. the Anglican Church, the Church of England). The curriculum merged seamlessly with what Charles had experienced at Shrewsbury: more Latin, more Greek, most of it learnt quickly, with the help of a paid tutor, simply to get through this or that exam.

Regular attendance at chapel was mandatory, and all the teaching staff were single men in 'holy orders', priests ordained by the Anglican church.

---

[1]Thomas Arnold, *Introductory Lectures on Modern History, Delivered in Lent Term, 1842*, 2nd edn (London: B. Fellowes, 1843).

When they married, they had to leave their college post. Often enough their college provided them with one of the many 'livings' they controlled. It was this endowed income (the 'living') which made the Church so appealing to fathers like Richard Darwin. With few responsibilities, a secure income, free housing in a pleasant rectory and high status within the parish community, Anglican livings were convenient places to park second sons who would never inherit much wealth and who lacked the intelligence or initiative to enter other walks of life.

Not all livings were the same, however. Some came with a larger salary than others. Many priests did not even live in their parish, but hired curates to carry out their duties to care for the souls of their parishioners. Livings could be enjoyed in plurality (i.e. one priest could hold several livings). The local elite usually controlled who was appointed to a particular living. The right to nominate individuals to this or that parish was a form of property in itself; landowning families (as well as Oxbridge colleges) collected them along with their various estates. Like any *ancien regime*, this pre-1832 world was built on patronage, on the giving, exchanging and withholding of favours between patrons and their dependents, their 'clients'.

Patronage networks were built on top of family relationships and reflected a world in which talent came second to birth. Every post, from the professorship of Botany at Oxford through a colonelcy in a guards regiment and a cathedral deanery to a post collecting customs and a seat in parliament – as we saw in the previous chapter, even jobs in training hospitals – young Darwin's world was ruled by patronage. The Ecclesiastical Revenue Commission (1834) and the Pluralities Act (1838) saw parliament legislate to reform ecclesiastical patronage. But this was seen as a case of removing 'abuses' rather than representing a move to disestablish the Church of England, that is to remove its privileged status as the state church, represented in the House of Lords (through its bishops) and funded by church rates paid by Anglican as well as non-Anglican residents of the parish. Charles was born into this world. He had little reason to challenge it.

At Cambridge young Charles would not have to work hard, but could be trained up in 'divinity' (theology) and perhaps befriend some young lord who might be able to give his college friend a living further down the line. Here again there was a rather more disciplined relative to go to for advice: Charles' cousin, William Darwin Fox, whose training for the Church was proceeding nicely. One did not need to feel much of a calling or 'vocation' to become a priest. Indeed, passionate expressions of faith were viewed with embarrassment, as regrettable signs of 'enthusiasm' (a pejorative term, associated with insanity). During the 1830s and 1840s the aforementioned attempts to reform the Church of England and to remove civil disabilities preventing non-Anglicans from participating in public life met with resistance from a band of young Oxford-based priests. Starting with Edward Pusey and continuing with the great John Henry Newman, the leaders of this so-called Oxford Movement challenged what they saw as 'national apostasy', a secularism they linked to the rise of utilitarian views of society and the state. Darwin escaped the violent intellectual upheavals that rocked the universities and the Church of England in these years, culminating in the shock of Newman's admission to the Church of Rome (i.e. Roman Catholicism) during the evening of 9 October 1845. Charles presumably shared the view that this fierce resistance to secularism was simply 'Papist enthusiasm' run wild. For his part Charles had little to say about church doctrine. If anything, that would have made him a safe choice for a fat Church of England living.

In his autobiography Charles Darwin would later write that his father's plan simply fizzled out over time, without there being a clear moment at which this fallback was explicitly abandoned. It would seem clear, however, that Charles' voyage round the world on a Royal Navy brig, *HMS Beagle*, marked a turning point. Charles was not in the Royal Navy and travelled as a gentleman companion to the brig's captain, Robert Fitzroy. Fitzroy had orders to complete a survey of the coast of South America, facilitating trade and helping to show the flag in a region of increasing importance to Britain, who had encouraged various 'liberators' (renegade officers, foreign mercenaries and other strong men, for the most part) to create independent nations in lands which had previously been Spanish or Portuguese colonies.

Fitzroy knew that this task would take 2 or 3 years, during which time he would have to ensure that the mask of command did not slip. There would be nobody he could talk to but his subordinates, who had to be kept in their place. The *Beagle*'s previous captain had buckled under this pressure, shooting himself. Though Fitzroy had then assumed command, he hoped that having a civilian companion would help him avoid a similar fate. This was particularly important as Fitzroy hoped to use his political influence (the nephew of a Duke, he, too, owed his position to patronage) to have the *Beagle*'s voyage extended into a circumnavigation of the globe. As it turned out the *Beagle* was away for 5 years and 2 days. According to the arrangement worked out by Fitzroy and Sir Francis Beaufort, hydrographer to the Navy, Fitzroy's companion would have a privileged position, dining with the captain. But he would have to pay his way: £500 in addition to his own personal costs of purchasing equipment and hiring horses and guides whenever he ventured onto land.

Charles Darwin was not the first to be offered this post. Three other candidates came before him, but bowed out for various reasons. Charles' qualifications were primarily social: he was a gentleman. Those influential people pushing his candidature also cited other, more 'scientific' qualifications. These qualifications were sufficient to overcome initial resistance from Robert Darwin. When Charles received the invitation in August 1831 and approached his father about it, Robert as well as Charles' sisters had many questions (the cost – a hefty £25,000 in today's values – was not much of an issue, apparently). Was the ship safe? What had the Admiralty seen in this 22-year old? What exactly qualified Charles Darwin to go on such a voyage?

As a schoolboy Darwin had collected birds eggs. Birds-nesting, like fishing, riding and hunting certainly encouraged him to spend time in the natural world. These were common enough hobbies for a boy of his rank, however. The chemical experiments he performed in a garden shed were unusual, but there he simply followed his older brother, Erasmus. The first indications that Charles had a taste for more specialized pursuits as well as for broader theoretical controversies appeared in the course of his otherwise unsuccessful time at Edinburgh University. In 1826 Charles joined a student society for natural history, the Plinian Society, which met regularly in a basement to listen to and discuss papers researched and written by the students themselves. The Society was open to the new phrenology as well as to the Lamarckian ideas imported by Grant. Darwin's interest in sea pens found in the nearby Firth of Forth led him to meet Grant himself and discuss his findings. He also attended lectures in zoology and geology by Robert Jameson, which, strangely enough (for the time) included hands-on practicals. An admirer of his grandfather Erasmus' work, Grant took Darwin to meetings of more adult scientific societies and encouraged him to read Lamarck as well as Erasmus Darwin's *Zoonomia*.

Though the encounter was brief, Grant was the first of what would be a series of mentors who talent-spotted Darwin and encouraged him to take his researches further: lending hard-to-find books, reading draft papers, discussing ideas and in some cases taking the young man on extended one-on-one field trips around different parts of the country. In contrast to the advanced ideas common at Edinburgh, at Cambridge Darwin fell among scholar-clerics, people like the Rev John Stephens Henslow, a botanist, and the Rev Adam Sedgwick, a geologist, who seemed happy enough with the Paleyite *ancien regime*. Henslow was the most important and secured the crucial invitation to travel on the *Beagle*.

Having initially been appointed to the chair in Mineralogy 2 years before, in 1825 Henslow secured a second chair in Botany, on the death of its previous holder, Thomas Martyn. Henslow confessed to knowing little of botany, but clearly felt he could do better than his predecessor, who had been appointed way back in 1762. Henslow did much to develop the university's botanic garden, which he felt Martyn had neglected. Yet Martyn had hardly been a nonentity and had himself struggled to interest students in natural history. In 1782 Martyn noted to a friend how rare an interest in 'Natural History' had been in Cambridge around 1750 when he had been an undergraduate. 'We were looked upon as no better than weed gatherers', he recalled, 'and I can remember very well that when I walked forth now and then, with a little hammer concealed under my coat, I looked carefully round me, let I should be detected in the ridiculous fact of knocking a poor stone to pieces.'[2] Eighty years on, Darwin's habit of 'weed gathering' and hiking out in search of rock samples still struck Cambridge contemporaries

[2]Thomas Martyn to Robert Strange, 8 April 1782. British Library, Egerton MSS 1970, f. 80.

as odd. Darwin's fellow students came to calling him 'the man who walks with Henslow'.

Darwin's interest in beetles was less unusual, being something of a fad at the time. Here Darwin was inspired by his cousin, Fox. The beetles were caught by hand, sometimes in the detritus left at the bottom of barges carrying reeds in from the Fens (Darwin paid someone else to climb into the barges). Identifying and mounting the beetles gave Darwin his first insight into the challenges of taxonomy, which were particularly acute in the case of the rich diversity of beetles within reach. He had to learn to compare printed descriptions in entomological handbooks and primers with his own observations. Darwin took great pride in seeing his name in print, credited with catching a rare beetle in an instalment of James Francis Stephens' *Illustrations of British Entomology*.

Fox of course was training for the church, and Henslow and Sedgwick were both ordained priests who eventually took a parish (Sedgwick held his alongside his professorial chair). Apart from the aforementioned appeal to his father, the Church promised Charles a job which would leave plenty of time to continue collecting beetles and geological specimens. There was no need to choose between the Church and science. Far from being opposed to each other, from young Darwin's perspective the two went together: as a country vicar, he would be able to accommodate activities alongside his official duties, activities which might seem eccentric or pointless on their own.

Among the books Henslow recommended to Darwin was Alexander von Humboldt's *Personal Narrative*, a detailed, seven-volume account of the Prussian's forays around Central and South America between 1799 and 1804. Humboldt had surveyed the course of the Orinoco River, braving electric eels and collecting data on climate and volcanoes as well as collecting many species new to science. A Romantic hero and *Naturphilosoph*, Humboldt's charisma appealed to Darwin as it had to Henslow, whose earlier hopes of visiting Africa had been in vain. Darwin planned an expedition to Tenerife, in the Canary Islands, and managed to talk Henslow and some other friends into agreeing to join him. He accompanied Sedgwick on a trip to map the geology of North Wales. This would, he hoped, offer him a useful crash course in geological fieldwork that would stand him in good stead in Tenerife. Returning home to Shrewsbury on 29 August 1831, he found a letter from Henslow, offering him the chance to venture on the *Beagle*.

As Henslow noted at the time of the offer, Darwin was not a '*finished* naturalist' in 1831. In 5 years, however, Darwin had cobbled together a working knowledge of geology, zoology, entomology and botany. Though Henslow noted that he needed to read more widely, Darwin had already learnt a good deal through hands-on practicals and fieldwork, as well as by many long conversations with men who held prestigious chairs. These were qualifications to place alongside his genteel background and his family's financial support. A final factor was timing. Whereas Henslow had had to back out of a much shorter trip to Tenerife after his wife gave birth to a child, Darwin had no ties. He was ready to go.

The following weeks were given over to feverish preparations. The staff of the British Museum did their best to fill in his gaps in knowledge, especially when it came to preserving and packing the specimens he would collect for the long voyage home. He bought a telescope, a compass, a rifle and a pair of good pistols to 'keep the natives . . . quiet'. Charles was exuberant, boyishly so, as his mind set sail to imaginary realms of adventure familiar from his reading of Humboldt. 'We shall have plenty of fighting with those d[amned] Cannibals,' he wrote to one school friend, adding that 'It would be something to shoot the King of the Cannibal Islands.'[3]

The voyage of *HMS Beagle*

The *Beagle* made it out of Devonport on the third attempt, giving Charles an early taste of the nausea he would experience throughout the voyage. The ship headed south, stopping at the Cape Verde Islands off the western coast of Africa before reaching Brazil in February 1832. The ship continued south, along the coast of Argentina and as far south as the Falkland Islands before arriving at the stormy southern tip of South America, Tierra del Fuego, in December. One of Fitzroy's secondary tasks was to return three Fuegians native to this inhospitable territory, who had been picked up and brought to Britain several years before. It was hoped that they, having adopted British dress and manners, might help to 'tame' their fellow Fuegians. A British missionary was to accompany them in this Christianizing, civilizing mission. The *Beagle* dropped them off, returning a month later to check on progress.

Darwin got to know the three Fuegians well before they reached their home, and considered them to be his equals in morphology and intellect. They were not a separate race. Yet he found their oddly unemotional reunion with their relations, their rapid return to native customs (which included behaviour totally unacceptable in Britain, then and now) and the treatment of the missionary (who was plundered, roughed up and soon had to be rescued) confusing and thought-provoking. The 'beneficiaries' of Britain's charitable mission seemed to have learnt nothing from their experience. If it wasn't some innate handicap, what else could explain their reasons for clinging to a hunter-gathering lifestyle in such a miserable, cold, wet corner of the continent, which even beetles seemed reluctant to inhabit? Though there would be little discussion of the Fuegians in Darwin's *Voyage of the Beagle*, the account of his voyage published in 1839, let alone in *The Origin*,

[3]Adrian Desmond and James Moore, *Darwin* (London: Michael Joseph, 1991), p. 104.

Darwin's encounter provoked questions that he chewed over for the rest of his life.

Darwin's travelogue, entitled *The Voyage of the Beagle*, closes with several pages of advice to readers considering a similar trip. Darwin writes as a seasoned traveller, and it is easy to assume that Charles knew exactly what he was doing on the *Beagle* and that the 'young naturalists' who read Darwin's *Voyage* and followed in his footsteps travelled in similar style. They did not. Born into a lower middle-class family that had been scattered to the winds by the collapse of his father's school in Ealing, Huxley had to travel in uniform, having joined the Navy in hopes of being able to pay back debts. Huxley served as assistant surgeon on *HMS Rattlesnake* between 1846 and 1850. Charles' social rank and wealth made it easy for him to disembark and spend anything from a few hours to a few months roaming around on land. Huxley had to obey orders and ask permission to go ashore, which wasn't always given.

Charles was free to return to Britain any time he wished, safe in the knowledge that he had a secure home and family waiting for him in Shrewsbury. Alfred Russel Wallace's background was similar to Huxley's. He funded his voyages by selling specimens to a dealer back in London, who in turn sold them to rich collectors who could not be troubled to travel and get their own hands dirty. Darwin could well have ended up as one of these armchair naturalists. Faced with his father's initial refusal to let him go on the *Beagle* Charles had not been overly troubled. A Paleyite did not question the Father's authority. Instead Darwin had made arrangements for the upcoming partridge-shooting season, which he was greatly looking forward to.

CHARLES LYELL'S *PRINCIPLES OF GEOLOGY*

Darwin's cabin on the *Beagle* was well stocked with books on natural history, none of them quite as well thumbed as the first volume of a text which has become a landmark in the history of geology: Charles Lyell's *Principles of Geology*. Although Darwin only met Lyell in person in October 1836, after his return, the *Principles* was a constant companion to Darwin on his travels, as well as a gateway to Lamarck, whom Lyell himself critiqued as misguided and overly speculative. The second volume of the *Principles*, which Darwin received on arriving in South America, acknowledges the species problem and considers how a 'student' might end up adopting Lamarckianism.

Lyell's key principle was uniformitarianism (the term itself, however, was coined by the philosopher William Whewell, a fellow of Trinity College, Cambridge). True geology could not admit any cause which we could not observe going on around us, could not summon up massive 'revolutions' in the earth's structure, even ones, such as the Flood, mentioned in holy scripture. That is not to say that Lyell's model was uneventful: if anything, earthquakes and volcanoes became business as usual, two of a number of interconnecting forces constantly at work around us, including, crucially, the effect of plants and animals on the earth.

'Tribes' of this or that creature went extinct, and new ones were 'called into existence' as the environment changed. Lyell invites us to imagine the changes such a tendency would bring about 'throughout myriads of future ages', not only on land, but also in the oceans.

> The mind is prepared by the contemplation of such future revolutions to look for the signs of others, of an analogous nature, in the monuments of the past. Instead of being astonished at the proofs there manifested of endless mutations in the animate world, they will appear to one who has thought profoundly on the fluctuations now in progress, to afford evidence in favour of the uniformity of the system, unless, indeed, we are precluded from speaking of *uniformity* when we characterize a principle of endless variation.[4]

Though 'mutations' here is not being used in a transmutationist sense, Lyell's destructive nature would certainly have shocked Paley, who had presented the Creator as careful not to lose any of his creatures to extinction. Yet the success of the *Principles* was not viewed as dangerous, in the way, say, the success of *Vestiges of Creation* would be. Why?

Lyell's format, tone and register were crucial in this regard. At 1,500 pages, the *Principles* was not a quick or an especially affordable read. It contained quotations from classical authors like Virgil and Pindar, authors familiar to the elite. Lyell's *Principles* did not present the earth as developing in any particular direction. If anything, it went in circles. If the climate of the Sussex weald were to become warm again, he suggested, the iguanodons might well reappear. It was foolish to look for progress. The vast scale of deep time encouraged humility, patience, veneration – Christian virtues, all.

By August 1833 the *Beagle* had moved back up the eastern coast of South America where Darwin spent several weeks travelling with gauchos, the wandering cowboys of the pampas. This was romantic adventure, Humboldt-style: days on horseback, nights under the stars, pistols at the ready. Though Darwin was an excellent shot, he was less of an action hero when it came to mastering the lassoo and the *bolas* (a throwing weapon consisting of balls connected by cords) with which the gauchos caught cattle as well as guanacos (a form of llama) and rheas (large emu-like birds). Though few people understood what it meant, Darwin's unofficial title of *señor naturalista* smoothed his passage through areas experiencing civil unrest. Darwin seems to have charmed General Juan Manuel de Rosas, a gaucho turned soldier then engaged in a low-level conflict on the inland frontier

[4]Charles Lyell, *Principles of Geology*, ed. James A. Secord (Harmondsworth: Penguin, 1997), p. 277.

of the Confederation of Argentina, intended to subdue the indigenous mapuche and protect ranchers' interests. Argentina's rocks yielded fossils of a giant armadillo (the Toxodon) and other large vertebrates for whom it was hard to find modern equivalents (the Megatherium, the Mylodon). 'How wonderfully are the different Orders, at the present time so well separated, blended together in different points of the structure of the Toxodon,' Darwin observed in the *Voyage*, as if delighting in the mess his finds were making of Cuvier's system.[5]

Darwin rejoined the *Beagle*, which now moved round to the western coast, avoiding the stormy open seas by passing through the Beagle Channel (which had earlier been named after the ship). When an earthquake measuring around 8.1 on the Richter scale struck near the coastal city of Concepción (levelling two thirds of it) in the afternoon of 20 February 1835, Darwin was safely located a couple of hundred miles to the south, enjoying a nap on the island of Chiloé. Darwin was close enough to be woken up, however, and to undergo its profound effects. To find something one had assumed to be solid (the ground under one's feet) undulating like rubber is to undergo a profound change in worldview, one which is hard to put into words. 'A space of a second was enough to awaken in the imagination a strange feeling of insecurity which hours of reflection would not have occurred,' Darwin later wrote.[6]

He recovered sufficiently over the following weeks to note how the upheaval had caused the whole landmass to rise (this 'uplift' reached 3m in some places), leaving mussels and other molluscs that relied on the tide to keep them moist and fed stranded 10 feet above the high-tide mark. This shock, accompanied with the associated tsunami and the volcanic eruption of nearby Mount Osorno provided ample food for speculation on how relatively routine forces could change the landscape and confront this or that creature with a new set of environmental challenges. In his *Voyage* Darwin transplanted the cataclysm to England, imagining the chaos that would result. 'England would at once be bankrupt'; he wrote, 'all papers, records, and accounts would from that moment be lost. Government being unable to collect the taxes, and failing to maintain its authority, the hand of violence and rapine would remain uncontrolled.'[7] To readers safe at home (which Providence had seen fit to locate somewhere far away from any fault lines), it was just the thing to send a pleasant chill down the spine.

Continuing north, the *Beagle* reached Valparaiso, where Darwin disembarked and went on another of his long forays into the interior, spending 4 months exploring the Andes, crossing over into Argentina and back. Here as elsewhere Darwin was not blazing new ground. In *The Voyage of the Beagle* he regularly compares his observations with those of earlier travellers

---

[5]Charles Darwin, *The Voyage of the Beagle* (Danbury, CT: Grolier, 1980), p. 89.
[6]Ibid., p. 306.
[7]Ibid., p. 309.

like Sir Francis Head, Humboldt and others. Even remote regions had the odd English rancher, consul or mining director, and Darwin secured letters of introduction to them in advance. His ample funds helped to hire guides and buy six mules to carry them all around. Though it was rough at times and Darwin occasionally fell ill, he never passed the edge of 'civilisation' as his co-discoverer Wallace would do, encountering peoples who had never seen a European before. Darwin did not see his ship consumed by fire on the high seas, as Wallace did, never found himself stranded in a lifeboat wondering if he would survive. Though Darwin bunked with two young 'middies' (midshipmen, trainee officers) in one of the *Beagle*'s two cabins, they treated him with respect, giving him the nickname 'Philos' ('Philosopher'). On *HMS Rattlesnake* Huxley's middies delighted in throwing his partly dissected specimens overboard when they started to smell, contributing to his bouts of severe depression.

The *Beagle* arrived in the Galapagos Islands off the western coast of Ecuador in September 1835 and remained in the archipelago just over a month. As noted in the introduction, the 13 islands had emerged from the sea as a result of volcanic activity, and the environment on each differed, with the 'older' islands more lush than the 'newer' ones. Different vegetation created different habitats and the finches and other creatures which had flown or been carried onto the islands from South America had evolved slightly different features to fit a range of niches. The islands were unpopulated by humans when a Spanish bishop happened on them in 1535. Having served as a haven for pirates and then, in the early nineteenth century, whalers, in 1832 the newly independent republic of Ecuador claimed them, sending out a governor and establishing a penal colony. Abortive attempts to sell them to the United States were followed by the creation in 1959 of a National Park. This laid the foundations of what has become an international tourist destination universally associated with Darwin, his finches and his discovery of natural selection. The reader of Darwin's *Voyage* can therefore be excused for feeling a certain disappointment on finding a mere 29 pages devoted to the Galapagos, out of a book of 509 pages. In a way, however, it underlines Darwin's self-identification as a geologist at this point in his life. His main aim was to study mountain ranges, volcanic craters and atolls for evidence of the subsidence and uplift described in Lyell's *Principles*.

Darwin's observation of differences between the morphology of mocking-birds, finches and other creatures as well as his counting of the number of unique species found on individual islands nonetheless sowed the seeds of a great discovery. Among the detailed descriptions of finch beaks, of the drinking habits of giant tortoises and the stomach contents of Amblyrhynchus lizards the following paragraph grabs our attention:

> The archipelago is a little world within itself, or rather a satellite attached to America, whence it has derived a few stray colonists, and has received the general character of its indigenous productions. Considering the

> small size of the islands, we feel the more astonished at the number of their aboriginal beings, and at their confined range. Seeing every height crowned with its crater, and the boundaries of most of the lava-streams still distinct, we are led to believe that within a period geologically recent the unbroken ocean was here spread out. Hence, both in space and time, we seem to be brought somewhat near to that great fact – that mystery of mysteries – the first appearance of new beings on this earth.[8]

These creatures could not have been placed on their islands 'in the beginning', because the islands weren't around then. After Concepción, Darwin would have found it more difficult to see them as the product of a rare, extraordinary rearrangement of vast continents and seas by God. The creatures he observed in the archipelago were not simply strays from the mainland, but distinct species. Had God waited for them to emerge from the deep and then created one finch for James Island, another for Chatham Island and so on? Darwin was too busy to think too hard about it, for now.

On 20 October the *Beagle* set off on its 3,200-mile crossing of the Pacific, barely pausing to look at passing atolls. In mid-November they reached Tahiti, which Darwin found impossibly, romantically lush after the Galapagos, drinking from coconuts and feasting on bananas as native guides led him and his new personal servant, Syms Covington, into the interior. New Zealand, the *Beagle*'s next stop, was a disappointment; the natives lacked the Tahitians' charm and the English were 'the refuse of society'. Darwin tallied up the four Christmases he had now spent away from home and looked forward to returning. Mid-January saw Darwin in Sydney, Australia. Here as in Tasmania off Australia's coast Darwin observed the curious, rapid decline of indigenous populations relative to English settlers.

In April 1836 the *Beagle* called at the Keeling or Cocos Islands, on the eastern edge of the Indian Ocean. In May it crossed over to Mauritius, famous as the home of the extinct dodo. Darwin considered the formation of atolls, proposing that they were formed as subsidence slowly submerged mountains or a coastline. As the land went down, the coral grew upwards, always keeping the top of the resulting reef in the sunlit upper level. Given the speed of coral formation, Darwin noted, this sinking 'must necessarily have been extremely slow'. The shape and level of reefs thus served as 'wonderful memorials of the subterranean oscillations of level', giving the geologist 'some insight into the great system by which the surface of this globe has been broken up, and land and water interchanged'.[9] Early essays on the 'Elevation of Patagonia' (mid-1834) and 'Coral Islands' (late 1835) saw Darwin building on but also challenging Lyell's geology, arguing that subsidence and uplifting were related.[10] These essays were later revised and

---

[8]Ibid., p. 382.

[9]Ibid., p. 485.

[10]Sandra Herbert, *Charles Darwin, Geologist* (London: Cornell University Press, 2005), p. 171.

expanded into books: *The Structure and Distribution of Coral Reefs* (1842), *Volcanic Islands* (1844) and *Geological Observations on South America* (1846).

Rounding the Cape of Good Hope in June, the *Beagle* entered the Atlantic, stopping at St Helena, where Napoleon had been exiled after Waterloo, as well as at Ascension Island. The ship arrived on the coast of Brazil on 1 August, completing its chronometrical measurements. Even as he yearned for home Darwin could not resist marvelling at the lush landscape of Brazil. Yet a week or so later we find Darwin thanking God (not 'Nature') that he would never have to visit a slave country again. In the *Voyage* Darwin briefly lists some of the 'revolting details' of humanity's cruelty towards fellow humans which he was forced to witness, in Pernambuco and elsewhere. Darwin's anti-slavery views clashed with those of Fitzroy, who claimed to have met slaves who were happy with their lot and who said as much when asked. Fitzroy became so angry in the course of one discussion of slavery that it had seemed likely that Darwin would have to leave the ship. Though Fitzroy apologized, Darwin alluded to this exchange in the *Voyage*, without, admittedly, going so far as to name the captain of the *Beagle* as one of those 'blinded' to slavery's horrors.[11]

Darwin left the *Beagle* at Falmouth on 2 October 1836, arriving at home in Shrewsbury 2 days later. His family had not expected him to arrive so early, and so his sudden appearance at the breakfast table surprised and delighted them. Robert Darwin immediately noticed that his son's head had changed shape (history does not relate which phrenological organs had been developed). It was clear that he had matured as well as found 'an interest for the rest of his life', as his sister Caroline put it.[12] There was no more talk of finding a fallback position for him in the church.

Mental rioting

Within weeks of his return Darwin was back with Henslow in Cambridge, who secured a £1,000 grant from the Treasury to fund the cataloguing of the specimens Darwin had brought back, as well as the publication of these findings in a 19-part *Zoology of the Voyage of HMS Beagle*. Darwin spent the next 3 years in Cambridge and London, busy cataloguing specimens, writing his *Voyage* (the non-technical account of his travels), meeting fellow naturalists and presenting his first scientific paper at the Geological Society. He became its secretary in 1838, joining the Athenaeum (a gentleman's club in London) the same year. He proposed to his cousin Emma, marrying her in early 1839. With her £5,000 dowry, £10,000 from his father and a promise

[11]Darwin, *Voyage of the Beagle*, pp. 501–2.
[12]F. Burkhardt and S. Smith (eds), *The Correspondence of Charles Darwin*, 7 vols (Cambridge: Cambridge University Press, 1985–1991), 1: 505.

of a further £400 a year thereafter, the couple would never experience want. At today's values, it was a lump sum of around £660,000, with an income of £17,000.

Before proposing Darwin did a cost/benefit analysis, noting, for example, that a wife could provide companionship ('better than a dog anyhow') and children ('if it Please God'), albeit at a cost: Darwin would not be able to travel and might have to put up with squabbling, while even charming 'female chit-chat' came with '*terrible loss of time*'.[13] Emma noticed that Darwin's religious beliefs were unconventional and may well have performed her own cost/benefit analysis. At 31, she was at risk of being left a spinster. Within a year of marrying Darwin displayed signs of a mysterious gastric illness which he seems to have picked up on his travels. Emma found herself nursing her husband as well as her first child. In 1842 Darwin moved his rapidly growing family from London to a house on the edge of Down, a sleepy village in Kent whose quiet would, he hoped, be good for his health. It does not seem to have helped. Nor did water cures at Ilkley and other spas. The 'vomitings' continued unabated.

The precise nature of Darwin's illness remains unclear. It is sometimes suggested that he played it up to secure the calm and space to think and write undisturbed, as an alibi whenever he might have faced uncongenial company or unwelcome invitations in London, which was well within reach – for a healthy person, at least. It is nonetheless clear that the costs in time lost to attacks of the illness outweighed the gains. As if to compensate, Darwin made the house and garden into a kind of laboratory for his researches, co-opting his children to help with the routine tasks involved in counting earthworms, caring for pigeons and catching insects to feed the carnivorous plants. Down House has been restored and is now open to the public. In addition to the sand walk on which Darwin took his daily walk one can admire the mirror he set up outside his study window, allowing him to inspect would-be visitors (and decide if he was willing to see them).

Floods of letters poured out of his study to London and to the farthest reaches of the world. Couched in polite, self-effacing language, these letters posed a series of questions which must have struck their recipients as odd. Would a female cat whose tail was cut off have tailless kittens? Were the fish in the Black Sea similar to fresh-water or salt-water fish? Was it true that a castrated cockerel would sit on eggs? Did tea trees degenerate when transplanted to Brazil? What on earth was Darwin driving at? These correspondents were having their pockets picked, in a way. They were supplying pieces to test 'my theory', a theory of selection none of them knew existed. Lyell did much the same. But whereas American colleagues came to call Lyell 'The Pump' (always pumping them for information), suspecting him of stealing their thunder, Darwin's disarming manner did not raise the same hackles.

The pieces were added to small notebooks in which Darwin began collecting his thoughts on transmutation. The key 'transmutation' notebooks

---

[13]Desmond and Moore, *Darwin*, p. 257.

date from 1837 to 1838 and are labelled B, C, D; those labelled M and N considered the origins of human behaviour. Alongside nuggets contained in correspondence these notebooks included notes taken from Darwin's wide reading as well as from long conversations with his father and other naturalists, but also with animal breeders. These breeders included individuals Lyell would probably not have deigned to address, including the unnamed young man at Willis', the barber in Great Marlborough Street, London, where Darwin had his hair cut.

'Notebooks' may summon up images of carefully kept, neat collections of data, tidy descriptions of carefully controlled experiments, perhaps with equations and tables full of numbers. The reality is far rougher, almost slapdash. In notebook C we find a typically dense passage of notes:

> At the end of "*White's Selbourne*." many references very good. also "*Rays Wisdom of God*". Often refer to these. – Also some few facts at end of "*The* British Aviary" or Bird Keepers Companion Study Appendix (& only appendix) of Congo Expedition NB. I met an old man –, who told me that the mules between canary birds & goldfinches differed considerably in their colour & appearance . . . . Tuckeys voyage – p. 36 "Cercopithecus sabœus" said to be monkey of St Jago C[ape] de Verd; same as on coast of Africa. – **Macleay tells me same thing**[14]

In this part of notebook C (C248–249e) Darwin seems to be considering how species are distributed across the globe. Though difficult to interpret, the range of Darwin's reading and information gathering and information checking is clear.

Here Darwin draws on travelogues such as J. K. Tuckey's 1818 *Narrative of an Expedition to Explore the River Zaire*, English natural history (Gilbert White's 1825 *Natural History and Antiquities of Selborne*), seventeenth-century natural theology (John Ray's 1692 *Wisdom of God Manifested in the Works of Creation*), handbooks for bird keepers as well as conversations with a zoologist (William MacLeay) and some old man he once met. At first glance it seems like something a scientist might scrawl on a napkin and then throw away. Yet changes in ink and marginalia show Darwin returning again and again (the passage in bold was added later). At points in his notebooks Darwin records his amusement at surprising or ludicrous claims (adding '!' or even '!!!'). Sometimes he changes tack, imagining objections to his theory. The command 'think' appears more than once. The notebooks were clearly very precious to Darwin.

Darwin had distributed his *Beagle* specimens carefully, in a way guaranteed to smooth his entry into learnt societies and to reach out to Owen and other leading lights of natural history. His finches went to the Zoological Society, where they were catalogued by John Gould. It was

[14]Paul H. Barrett (ed.), *Charles Darwin's Notebooks, 1836–1844* (Cambridge: Cambridge University Press, 1987), pp. 248–9e.

Gould who noticed the famous variation in beaks – of the mockingbird specimens, not the finches. Darwin had failed to label his finch specimens by island, making it impossible to verify if they represented evidence of adaptation. Had Darwin's servant on the voyage, Syms Covington, not been more precise in his own record keeping, the finches would have missed their starring role in the history of discovery.

Darwin's reading of Malthus' *Principle of Population* in September 1838 led him to consider ruthless competition and demographics. Darwin used the phrase 'mental rioting' to describe the troubling, exhilarating process of speculation on what he called 'my theory', which only grew more riotous as he pursued his eccentric research into artificial selection (i.e. the breeding of domesticated animals), which he would later present in *The Origin* as a key analogy to the speciation he detected in wild creatures. In the E notebook Darwin recorded his reaction to William Whewell's recently published *History of the Inductive Sciences* (1837), which had argued against a law-driven 'hypothesis of progressive tendencies'. Already Darwin had an idea of what he was looking for: a non-teleological form of transmutation which would make variation a bridge to new species, rather than a sterile dead end. 'See if any law can be made out,' he wrote in his notebook, 'that varieties are generally additive, & not abortive'[15]:

In May 1842 Darwin wrote a 35-page sketch outlining how 'natural selection' chose the winners from the Malthusian struggle. Rather than racing one another up a rising Lamarckian path, living creatures were inter-related, with common ancestors. This in turn enabled a truer, genealogical classification. This was, he wrote, a grander vision of God, and one which could explain parasitism and extinction as tending towards 'the highest good, which we can conceive, the creation of the higher animals'.[16] Teleology crept back in. Though he kept the sketch private, a recent publication was making such ideas seem less shocking.

VESTIGES OF CREATION

Born in Peebles, Robert Chambers' family moved to Edinburgh in 1813 after his father's cotton spinning works went bankrupt. Chambers revelled in the historical novels of Sir Walter Scott as well as in the phrenology of George Combe. Unable to afford university he started his own bookselling business, publishing a radical newspaper named *The Patriot*. In the 1830s Chambers joined the 'useful knowledge' bandwagon; *Chambers Edinburgh Journal* and *Chambers Educational Course* offered accessible introductions to a range of historical as well as natural historical topics, appealing to working-class autodidacts as well as to respectable middle-class families.

[15]Barrett (ed.), *Darwin's Notebooks*, E70.

[16]Desmond and Moore, *Darwin*, p. 294.

DETECTIVE NOTES

Suspects

Suspects					
Col. Mustard					
Prof. Plum					
Mr. Green					
Mrs. Peacock					
Miss Scarlett					
Mrs. White					

Weapons

Weapons					
Knife					
Candlestick					
Revolver					
Rope					
Lead Pipe					
Wrench					

Rooms

Rooms					
Hall					
Lounge					
Dining Room					
Kitchen					
Ball Room					
Conservatory					
Billiard Room					
Library					
Study					

ral History of Creation and, more importantly, *estiges* began with an dust and gas (nebulae) ary systems, each with d from ideas developed 327) late in the previous side out, advancing from vest. Animals and plants nion. 'Development' (i.e. ambers insisted that his more traditional natural ntervened many times to troy this or that landmass

n embryology, Macleay's osse's 1836 experiments, ubbed *Acarus crossii*) by ew on Charles Babbage's how an apparently closed numbers. It was hard to dary laws of 'variety' and ndeniable, equally obvious, he chose.

a simple (if vague) thesis hat in speculating on such Chambers' hands even the npulses in the brain could

ent – a little mass which, t in its living constitution, Visdom, how admirable its n of that Power by which is

o publish *Vestiges* under his dition of 1884 (although his ng mystery lent *Vestiges* an parlour game. Its anonymity, rwin, Whewell, Hershel and to critique *Vestiges*.

[17]Ibid., pp. 153–4.

Criticism of *Vestiges* was strident, yet curiously ineffective at the same time. In public men of science pointed to factual errors (which Chambers corrected in later editions) and speculated on the harm *Vestiges* might do if it got into the hands of working men, women or other weak-minded people. Neither response tackled the evolutionary theory head-on. Privately one senses they were embarrassed. They recognized that the tidy Cuvierite plan of the Bridgewater Treatises could no longer be stretched to explain everything, but they preferred to ignore this problem and focus on the everyday scientific work of species – and strata – classification. The public's appetite for *Vestiges* indicated that this strategy – if that is in fact what it was – was no longer sustainable. The ideas contained in the work were controversial, but common enough among the educated elite for there to be 60 named candidates for its authorship. Chief among them was a Tory squire, Richard Vyvyan, known to have published similar views in his two-volume 1844 work *The Harmony of the Comprehensible World*.

In contrast to this host of possible authors the number of writers willing to rebut its arguments at length was few. Owen refused two invitations to review it, but agreed to help correct later editions when Chambers approached him (anonymously, through a third party). Although Whewell published a work entitled *Indications of a Creator* (1845), otherwise it was left to a stonemason-turned-geologist, Hugh Miller, to tackle *Vestiges*. Miller's *Foot-Prints of the Creator* appeared in 1849 and argued that the presence of highly developed creatures in lower geological strata contradicted Chambers' view of life developing from simple to ever more complex forms. Rather than being wiped out, God had made the dinosaurs smaller in size (iguanodon to iguana, as it were) so as to enable mammals to grow larger without fear of finding the earth 'a place of torment'.[18]

A far greater gulf yawned between Darwin's public and private personas. Darwin's notebooks show him ridiculing Whewell, Owen and others' views in private. But he did not do so in public. When the leading lights of the Geological Society plotted to destroy Grant's reputation, Darwin (as secretary) was happy enough to go along with them. Darwin may have heard rumours of Grant's homosexuality, something then viewed with disgust. Either way, Darwin's silence suggested that a concern for respectability trumped the gratitude he owed Grant, as one of the first men of science to take him seriously. That a man favoured by birth and fortune, with an established reputation as a serious natural historian thought it prudent to stay silent – to leave Grant to his fate, and, more importantly, to keep his own speculations a secret for two decades – indicates just how powerful that burden of respectability could be.

Darwin's silence also reflects a laudable desire not to come out in favour of transmutation before he had marshalled evidence that not only showed it happened but also illustrated the mechanism by which he proposed to

[18]Hugh Miller, *Foot-Prints of the Creator*, 2nd edn (London: Johnstone and Hunter, 1849), p. 295.

explain it. Three favoured individuals were let into the secret. In January 1842 Darwin told Lyell of the direction his speculations had taken, asking him to keep it to himself. In late 1843 he told the botanist Joseph Hooker who had just returned from his voyage to the Antarctic on *HMS Erebus*. 'I am almost convinced (quite contrary to the opinion I started with) that species are not . . . immutable,' he wrote.[19] Over the winter he drafted a much longer (231-page) sketch of his theory, which he completed in February 1844. He showed this to Hooker. Asa Gray (1810–88), professor of Natural History at Harvard University, was another confidant, initiated into Darwin's secret by letter in 1857. Meanwhile Darwin's own wife remained in the dark. Darwin added a note ordering her to publish the 1844 sketch, and placed both in a sealed envelope, to be opened only in the event of his death. In 1847 he attended the Oxford meeting of the British Association for the Advancement of Science, happy enough to pretend that he agreed with Bishop of Oxford Samuel Wilberforce's sermon to the assembled scholars, in which he ridiculed *Vestiges*-style speculation as half-witted.

Four years later his beloved daughter Annie died while he was taking a water cure at Malvern. As with Huxley, this episode marked a crisis in Darwin's relationship with Christianity. But still, he kept his theories to himself, and packed his sons off to Rugby, a public school. From 1847 on he immersed himself in the study of cirripaedia (barnacles), a project which ended up taking 5 years, far longer than Darwin originally intended, precisely because of the diversity Darwin encountered. He also began a curious relationship with another member of the rising younger generation, namely Huxley. Huxley did not join the trusted coterie, but nonetheless came to adopt an important role in Darwin's thought before *The Origin*, ironically as a 'bugbear' or bogeyman, someone whose scepticism towards transmutation made him a useful intellectual sparring partner.

On 18 June 1858 Darwin received a letter from Wallace in Malaya, sharing a theory he had recently come up with. Wallace had been publishing papers for several years, and Lyell had warned Darwin of the risk that Darwin would find himself overtaken by the younger man, whose speculations were clearly following a similar line. Though they had apparently bumped into each other in the same species tangle, Wallace had come at the problem from a very different direction.

Alfred Russel Wallace

When *Vestiges* appeared Wallace was a 21-year-old trainee railway surveyor living with his brother in Neath. Like Huxley's, Wallace's was a large, downwardly mobile middle-class family, one which bankruptcy had

[19]Charles Darwin to Joseph Hooker, 11 January 1844. Darwin Correspondence Database, http://www.darwinproject.ac.uk 729 (accessed on 2 May 2013).

scattered across Britain and the United States. Young Wallace had only a few years of formal education in Hertford Grammar School before being packed off in 1836 to live with his brothers. He spent some time in London, with a brother training to be carpenter in a builder's yard. He then went to live with another brother, a trainee land surveyor of no fixed abode. Both resented having to keep their younger brother in decent clothes. To keep him out of trouble they recommended that he spend time in this or that Mechanics Institute, reading books and listening to lectures intended to provide working-class men with 'useful knowledge'.

Alfred's eldest brother William was in demand, thanks to enclosure and the railway boom that were profoundly altering the English countryside. An enclosure act gave the wealthier members of a parish permission to divide commonly held parish land (commons) and distribute it among themselves and other ratepayers. For generations the common had helped the poorest in the parish to eke out a living without becoming fully dependent on wage labour. The commons provided much more than a source of firewood, a place to play and graze one's animals; they represented a common wealth, a social safety net free from the harsh conditions of the New Poor Laws. A labourer with a cow on the common had more bargaining power with his employer than one without such a resource. Whereas the latter's choice was simple (work at whatever wage was being offered, go to the workhouse, starve), the former left room for manoeuvre (one could slaughter the cow, eat or sell the meat and await a better offer – which usually came, as farmers could not leave crops to spoil unharvested in the field).

Surveyors were needed to map common lands preparatory to their being divided up, as well as to mark out the route of new railway lines, bridges, tunnels and cuttings. The railways' effects on rural communities were almost as profound as that of enclosure, partly because the 1840s boom in their construction was so sudden, whereas enclosure had been going on, here and there, for decades. Rapid transit changed mental and economic maps: this town was now closer (in terms of the time required to reach it) than that. Markets were closer (the marriage market, too, became bigger). Rural migration to town accelerated. In the case of Alfred as well as another future evolutionist, Herbert Spencer, the experience of surveying not only provided a grounding in geology and natural history, but also led them to think about how environmental factors changed human beings and the way their communities were organized. Together with his reading of works by the 1790s radical Tom Paine and a contemporary socialist, Robert Owen (not to be confused with the comparative anatomist, Richard Owen), these experiences led Alfred to question his parents' Anglican beliefs as well as the Paleyite consensus which saw inequality as something simply to be accepted.

Wallace knew a brewer's clerk named Henry Bates who shared his interests in geology, beetles and botany. Though Bates didn't live nearby, he was the only person with whom Wallace could discuss quinarianism, phrenology and of course *Vestiges*. In December 1845 Wallace wrote to Bates defending *Vestiges* against the latter's criticism. It was 'an ingenious hypothesis strongly supported by some striking facts and analogies but which remains to be proved by more facts', he wrote. *Vestiges* 'furnishes both an incitement to the collection of facts and an object to which to apply them when collected'.[20] In September 1847 Alfred's sister, Fanny, returned from working in the United States and took her brothers on a trip to Paris. On his return Alfred wrote to Bates, describing the boulevards and the Muséum national d'Histoire naturelle. He also recommended Owen's edition of Oken's *Naturphilosophie*, though he noted how hard it was to get a copy (at least, for a man of limited means).

Reading Oken, Humboldt, Darwin's *Voyage* (twice) and visiting Paris had left Wallace feeling 'rather dissatisfied with a mere local collection – little is to be learnt by it'. 'I sh[oul]d like to take some one family, to study thoroughly – principally with a view to the theory of the origin of species.'[21] Wallace persuaded Bates to join him on what was, given their reduced circumstances, a reckless endeavour: a self-funded expedition to the Amazon. The pair visited the British Museum in London to learn what they could about Amazonian insects and other creatures. William Henry Edwards, author of *A Voyage Up the River Amazon*, provided letters of introduction; William Hooker of Kew Gardens helped with passports. Most importantly, a dealer in exotic butterflies and insects, Samuel Stevens, agreed to an arrangement by which they would be paid for sending specimens (3p for each insect) back to London. In April 1848 the pair sailed on the *Mischief*.

By the end of May they were exploring the lush jungle around Pará. After a quarrel with Bates in July 1849 Wallace headed up the Rio Negro alone, apart from a couple of native guides. He observed that the river acted as a border between species which were similar, but not identical. In December 1849 Wallace wrote the first of a flood of papers he would send back to London from his travels, for publication in learnt journals. He also sent back thousands of birds, eggs and specimens to Stevens. Unfortunately the ship carrying him home, the *Helen*, caught fire at sea. Wallace was only able to take a tin box of notes with him in the lifeboat. Happily he was picked up and arrived safely back in Deal. In London Wallace attended meetings of the Linnean, Entomological and Royal Geographical Societies. He published his own *Narrative of Travels* in 1853. He even bumped into Darwin in the Insect Room of the British Museum. But it was clear that the loss of his notes and valuable specimens meant Wallace would have to head out again. In April 1854 he arrived in Singapore, for what would be an 8-year tour of the Malay archipelago.

---

[20]A. R. Wallace to Henry Bates, 28 December 1845. Natural History Museum, London. Wallace Papers, WP1/3/17.

[21]A. R. Wallace to Henry Bates, 11 October 1847. Natural History Museum, London. Wallace Papers, WP1/3/19.

Wallace found a useful patron in 'Rajah' Brooke, a British adventurer who had managed to appoint himself *de facto* king of the island of Borneo. Brooke provided a house which served Wallace as a base for tracking, collecting, identifying and labelling thousands of moths, beetles, birds, fish and other creatures, among them the orangoutang. A poor shot, Wallace tended to wing rather than kill the orangs he aimed at, giving them time to escape or die, out of reach, in the trees. Wallace nursed an orphaned orang for 3 months. Elsewhere Wallace benefitted from well-disposed British and Dutch miners, rubber tappers and coffee growers. But in travelling to the Aru Islands he also crossed the invisible line separating such outposts of empire and 'civilisation', entering regions untouched by European man. His adventures made his 1869 travelogue a best-seller, and today *The Malay Archipelago* (which is dedicated to Darwin) has an appeal which the somewhat stuffy *Voyage of the Beagle* does not.

But that was writing for a non-professional audience, a means of making money. Wallace's contributions to learnt journals from 1855 onwards were equally striking, however, for the way they challenged natural historians' notion of species. He had opened his own notebook on species change in March of that year, in preparation for a book tentatively titled 'On the Law of Organic Change'.

In September 1855 Wallace's article 'On the law which has regulated the introduction of new species' appeared in *Annals of Natural History*. In it he proposed that similar species had descended from a common ancestor or 'antitype'. The following May Wallace crossed the strait separating Bali from Lombok. He noted that the species on the Lombok side looked more like creatures found in Australia, while Bali's creatures were more like those of Asia. This short, deep strait marked a biogeographical border (later known as 'Wallace's Line'). The striking differences were evidence that Asia and Australia had been isolated from each other for millions of years, each evolving its own particular flora and fauna. Even after uplift had made them neighbours, evidence of this long evolutionary separation was obvious. It was this article which led Lyell to warn Darwin and to begin his own notebook on the species problem. Darwin wrote to Wallace, making a point of mentioning that he, Darwin, had been reflecting on such issues for 20 years, and claiming (incorrectly) that he had 'many chapters' of a book already in hand. Wallace continued to write and send off papers. In February 1858, during a 2-hour 'cold fit' brought on by malaria, he hit upon the mechanism that could explain the patterns he identified in the geographic distribution of life. He wrote his famous letter to Darwin.

Darwin was torn by conflicting emotions: relief at finding independent confirmation, as it were, of his ideas; wonder at how closely Wallace's theory came to his own (not just the ideas, but also the terminology chosen); disappointment at losing 'my originality'. A lesser man might have destroyed the letter and rushed into print. Letters sent from such a remote outpost went missing regularly enough, and Wallace was several thousand miles and many months away. Unable to cope and distracted by fears that a local outbreak of scarlet fever might strike his young family, Darwin handed the matter over to Lyell and Hooker. They came up with the idea of reading two papers (one by Darwin, one by Wallace) at a meeting of the Linnean Society on 1 July 1858. Both men were absent. Darwin's paper (cobbled together from the 1844 sketch and the 1857 letter to Gray) was read first, given the fact that he could point to 20 years' work in the field.

The response at the meeting was, as we have seen, muted. Wallace was happy to give his retrospective approval to the decisions taken, when he eventually heard of them. Interestingly, he did not rush home, either to share Darwin's glory or to make a case that it should have been his exclusively. Wallace did not consider himself the victim of a conspiracy, either in 1858 or later. Instead he joined the ranks of those intimates (Lyell, Hooker, Gray and Huxley chiefest among them) eagerly awaiting Darwin's book. This would not be the 'Species Book', that treatise which Darwin had begun to write. It would be something else. But what?

Further reading

Barrett, Paul H. (ed.), *Charles Darwin's Notebooks, 1836–1844* (Cambridge: Cambridge University Press, 1987).

Desmond, Adrian and James Moore, *Darwin* (London: Michael Joseph, 1991).

Herbert, Sandra, *Charles Darwin, Geologist* (London: Cornell University Press, 2005).

Kohn, David and T. F. Glick (eds), *Charles Darwin on Evolution: The Development of the Theory of Natural Selection* (Princeton: Princeton University Press, 1996).

Moore, James and Adrian Desmond, *Darwin's Sacred Cause: Race, Slavery and the Quest for Human Origins* (London: Penguin, 2010).

Ospovat, Dov, *The Development of Darwin's Theory. Natural History, Natural Theology and Natural Selection, 1838–1859* (Cambridge: Cambridge University Press, 1995).

Richards, Ruth J., 'Why Darwin Delayed, Or Interesting Problems and Models in the History of Science,' *Journal of the History of the Behavioural Sciences* 19 (1983): 45–53.

Secord, James A., *Victorian Sensation: The Extraordinary Publication, Reception and Secret Authorship of 'Vestiges of the Natural History of Creation'* (Chicago: University of Chicago Press, 2001).

Topham, Jonathan R., 'Beyond the "Common Context": The Production and Reading of the Bridgewater Treatises,' *Isis* 89 (1998): 233–62.

van Whye, John., 'Mind the Gap: Did Darwin Avoid Publishing His Theory for Many Years?,' *Notes and Records of the Royal Society* 61 (2007): 177–205.

Online Resources

Darwin Correspondence Project. http://www.darwinproject.ac.uk/

Darwin Online. http://darwin-online.org.uk/

CHAPTER FOUR

Reading *The Origin*

Darwin began writing on 20 July 1858, a couple of weeks after his and Wallace's papers had been presented at the Linnean Society. His original plan was to dash off a summary of the scientific treatise he planned to write, fully referenced and footnoted, that 'Species Book' tentatively titled *Natural Selection* – a work which only saw print in 1975.[1] Within a month it was clear that his pen was running away with him, despite setbacks due to illness and accident. Darwin was sending draft hand-written chapters to Hooker as he went along. Hooker misplaced one manuscript, which ended up being used as drawing paper by Hooker's children, necessitating a rewrite. Lyell helped Darwin find a publisher, John Murray, who published *The Principles of Geology*. Now a book, Darwin's work would carry the title *An Abstract of an Essay on the Origin of Species and Varieties through Natural Selection*. Given the uninspiring title, Murray initially planned a run of just 500 copies. When Darwin agreed to shorten the title to *On the Origin of Species by Means of Natural Selection*, he upped it to 1250. The book appeared on 22 November 1859 and cost 14 shillings (£30 today). It sold out immediately. Darwin sent complimentary copies to Henslow, Gray, Owen and many others. He described his book as 'one long argument', but the notes he sent with these copies were typically apologetic ('I may be egregiously wrong').

'One long argument'

The Origin opened in a similarly apologetic vein, referring to the work as an 'abstract', published prematurely, because of the author's awareness that

[1]R. C. Stauffer, *Charles Darwin's Natural Selection* (Cambridge: Cambridge University Press, 1975).

his poor health might rob him of the time to refine it. Yet the reader is carried along by the argument through Darwin's 14 chapters. The range of examples was broad. In one page we pass from a discussion of the 'electric organs' possessed by ancient fish (and lost, apparently, by their descendants) to the giraffe's tail which 'looks like an artificially constructed fly-flapper'.[2] Unusually for a work of Victorian zoology, the flow of the text is not interrupted by footnotes. The reader does not have to thread his way around these textual ladders, is not invited to dart downstairs into the clanking boiler-room of scholarship and check that all is ship-shape. Darwin did not provide sources for his quotations or give a bibliography, although there was one important illustration, a tree-like diagram illustrating 'Divergence of Taxa'. Darwin found this tree 'simile' useful: 'At each period of growth all the growing twigs have tried to branch out on all sides, and to overtop and kill the surrounding twigs and branches,' he wrote, 'in the same manner as species and groups of species have tried to overmaster other species in the great battle for life.'[3]

Darwin began the argument by considering 'variation under domestication', showing how breeders of cattle, sheep and pigeons can bring about changes in their animals by selective mating. As he noted, investigating this had taken him to places naturalists of his day rarely went, such as pigeon clubs, where pigeon breeders or fanciers competed for prizes for the finest specimen of a jacobin or tumbler (two varieties of pigeon). Though it was universally accepted that all such varieties descended from one species, the Rock Pigeon (*Colomba livia*), were anyone to present these varieties to an unsuspecting ornithologist as wild specimens the ornithologist would not hesitate to classify them as different species. Before allowing them to breed fanciers observed their birds for minute mutations, selecting (or not selecting) this or that mutation in a process which, carried over many generations of pigeon, would indeed justify the famed Sir John Sebright's boast that he could breed any shape pigeon you could think of. For the most part, however, Darwin sought to present this artificial (i.e. man-made) selection as unconscious; nobody knew the origin of this or that breed; breeders, like connoisseurs of painting, could never precisely define what made for a prize animal. Though they claimed to keep a breed pure, history showed that they had in fact created a range of diverse, artificial breeds.

Were these breeds, varieties or races distinct species, however? In Chapter 2 Darwin used arguments already familiar from the introduction to challenge the systematists who used circular arguments to distinguish between those 'important' organs (those that did not vary among species of the same order or genus and that therefore served to characterize it) and other, secondary ones (those that did vary, allowing two species from the same order or genus

---

[2]Charles Darwin, *On the Origin of Species By Means of Natural Selection* (Mineola, NY: Dover, 2006), pp. 122–3.
[3]Ibid., p. 81.

to be distinguished). The only guide to 'whether a form should be ranked as a species or a variety' seemed to be the consensus of opinion among 'naturalists': yet a comparison of floras (catalogues of a region's endemic plants) published in different countries showed that such a consensus was lacking. Instead of helping to plug gaps, terms like 'intermediate forms' only clouded the matter further.[4] The term 'species' was, he concluded, a label given 'for the sake of convenience', by naturalists struggling to contain nature's tendency to vary, to cause mutations.

In the next chapter Darwin considered how superfecundity created a Malthusian 'struggle for existence'. This struggle would 'select' between those mutations found to be beneficial to the creature, enabling it to survive and reproduce successively, and other mutations found to be of no significance, or positively harmful. The offspring of the form with the beneficial mutation would tend to inherit it and so have a better chance of surviving and reproducing. This principle, 'by which each slight variation, if useful, is preserved', was christened 'Natural Selection, in order to mark its relation to man's power of selection'. Natural selection acts constantly, unlike artificial selection, and produced far greater results.[5]

Throughout *The Origin* Darwin was careful to avoid language suggesting that the winners in this constant battle are fitter on some universal scale leading from lower to higher forms. 'Absolute perfection' was never reached. The 'perfection' (fitness, in other words) of any creature was perfection 'only in relation to the degree of perfection of their associates' (i.e. its predators, its prey, any other creature, in fact, inhabiting the same environment). Fitness was a moving target, therefore.[6] Trifling differences could be the deciding factor that condemned a species to extinction. Such differences were so small that we overlooked or even ignored them. Instead of being tiny and incremental the changes we imagined were huge, global cataclysms. Such catastrophist histories of the earth fed a smug pride in our own existence. 'So profound is our ignorance, and so high our presumption,' Darwin wrote, 'that we marvel when we hear of the extinction of an organic being; and as we do not see the cause, we invoke cataclysms to desolate the world, or invent laws on the duration of the forms of life!'[7]

In Chapter 4 Darwin introduced a second principle of transmutation: sexual selection, by which male life forms competed against other males to breed with females. Sexual selection empowered females to do their own selecting. Darwin was a man of his time, a time when women were denied the vote, were unable to attend university and were presumed to be physically as well as mentally inferior to men. It was hard, therefore, to imagine women having this power. Darwin wrote that 'it may appear

[4]Ibid., p. 31.
[5]Ibid., p. 40.
[6]Ibid., p. 296.
[7]Ibid., p. 47.

childish to attribute any effect to such apparently weak means'. Although he put off further discussion, he found no reason to doubt that, say, female birds could produce 'a marked effect' by selecting 'the most melodious or beautiful birds, according to their standard of beauty'.[8]

One of the important gaps in *The Origin* was a coherent mechanism to explain where the mutations come from, the mutations that artificial, natural and sexual selection required to do their work. At the start of Chapter 5 Darwin insisted that, contrary to what he may have implied earlier in the book, these 'variations' were not the product of chance. Instead he noted that 'the reproductive system was eminently susceptible to changes in the conditions of life', and concluded that mutations in offspring mainly resulted from 'this system being functionally disturbed in the parents'.[9] Darwin admitted that it was very hard to determine why, given this 'disturbance', this or that change occurred in the young. Darwin threw everything he could at this problem, including Lamarckian 'use-disuse'. Creatures that move into caves might lose their eyes because that organ is not being exercised. Over generations this organ would atrophy, in obedience to the aforementioned Lamarckian use-it-or-lose-it principle. Darwin gave a greater role to 'reversion', a kind of reverse evolution in which a form adopts the features of a distant ancestor. Darwin proposed that this reversion tendency existed in constant tension with an opposing tendency towards 'variation' (i.e. mutation). One is reminded of Cuvier's notion that life forms are subject to a kind of gravitational 'pull' towards 'their' type.

Though *The Origin* can be seen as a piling up of geological, biological and zoological facts which can be explained by natural selection (vestigial organs, say), Darwin wanted us to go further. There is a sense in which *The Origin* seduces us by its rhetorical strategy. We become co-conspirators. By Chapter 6, apparently, we are in too deep. If we have followed him to this point, we are committed to go all the way, even as Darwin tackles one of Paley's most cherished examples of God's intelligent design, the human eye:

> He who will go thus far, if he find on finishing this treatise that large bodies of facts, otherwise inexplicable, can be explained by the theory of descent, ought not to hesitate to go further, and to admit that a structure even as perfect as the eye of an eagle might be formed by natural selection, although in this case he does not know any of the transitional grades. His reason ought to conquer his imagination; though I have felt the difficulty far too keenly to be surprised at any degree of hesitation in extending the principle of natural selection to such startling lengths.[10]

Darwin duly sketched out how a compound eye might have gradually emerged by natural selection from the simplest form: a nerve coated with

[8]Ibid., pp. 56–7.
[9]Ibid., p. 83.
[10]Ibid., p. 119.

pigment and covered by a membrane of some kind. The same morphological feature could serve different functions at different points in its development, helping to bridge any apparent gaps.

The following chapters extended natural selection to non-morphological features, to explain instinctive behaviour, including the decidedly unpleasant varieties exhibited by certain ant species, which enslave other species and make them collect food for them. Chapter 8 was defensive, aimed at those, notably Huxley, who wanted to make intersterility (i.e. the inability of two life forms to reproduce sexually) into a clear barrier against transmutation. Darwin noted the difficulties in making neat distinctions: rather than a tidy wall (unions within the species fertile, those 'across the wall' separating it from other species infertile) interfertility/intersterility was a spectrum.

Only in Chapter 9 did Darwin come to the beginning, to the fossil record, the place a geologist like himself might have been expected to set out from. Chapters 9 and 10 considered the palaeontological evidence against transmutation, while Chapters 11 and 12 turned to the evidence of biogeography, the geographic distribution of life forms. Darwin made the important point that, in searching for 'intermediate' forms (i.e. the ancestor A, which creatures B and C both descend from) we should not expect to find a character (a set of morphological traits) located halfway between the character of B and that of C. The series was not BAC or CAB. A might well differ more from B and C than B and C differed from each other.

Opponents of transmutation then as now pointed to events such as the Cambrian Explosion (as we call it today) and other, similar cases, in which the geological record seemed to act in a non-uniformitarian manner: moving abruptly from a relatively lifeless stratum to one teeming with evidence of abundant life. Darwin fought for his theory by pointing to recent discoveries of whale bones in the upper greensand. Whereas previously Cuvier and others had insisted that whales appeared as part of a 'revolution' in the secondary period, this new evidence showed that they were around long before. 'Revolutions' were, Darwin suggested, non-events. It was another case of presumption coming up with facts founded on ignorance – or rather, on a very partial knowledge, of a very small fragment of the earth's history.

The earth was, Darwin estimated, around 300 million years old. In that vast age the earth's crust had experienced many changes. There were bound to be gaps, therefore, in the geological record. In the 1850s Darwin could not of course use a cinematic metaphor. But he comes close in pointing out that in viewing fossil evidence we were only viewing 'an occasional scene [a few frames in the film, as we might put it] taken almost at hazard, in a slowly changing drama'.[11] Though we would never be able to compile a complete fossil record, new data such as the whale fossil was leading us away from catastrophism, towards uniformitarianism and transmutation.

---

[11]Ibid., p. 198.

Turning to the distribution of life forms, Darwin considered the debate over whether life had been created at one point on the globe or whether there had been centres of creation scattered across the globe. Here he cited Lyell's work, Wallace's 1855 paper, as well as the results of his own experiments at Down. There Darwin had immersed seeds in salt water for various lengths of time, then planted them to see if they would germinate. Out of a total of 87 different kinds of seeds, 64 germinated after a month's immersion, suggesting that the oceans could well have transported seeds carried into them by floods across from one landmass to another, enabling plants to colonize new territories.

At the Zoological Society Darwin carried out similar experiments into the dispersal of seeds by birds. 'I forced many kinds of seeds into the stomachs of dead fish,' he reported, 'and then gave their bodies to fishing-eagles, storks, and pelicans; these birds after an interval of many hours, either rejected the seeds in pellets or passed them in their excrement; and several of these seeds retained their power of germination.'[12] The zoo at Regents Park presumably offered enough to keep Darwin entertained as he waited for the pelicans to defecate, so that he could bring the product back to Down for planting in the garden. His conclusion was clear. Although there were cases of plant or animal distribution which defied easy explanation, 'nevertheless the simplicity of the view that each species was first produced within a single region captivates the mind,' he wrote. 'He who rejects it, rejects the *vera causa* of ordinary generation with subsequent migration, and calls in the agency of a miracle.'[13]

Chapter 13 saw Darwin turn to embryology. Although he said little about recapitulation, Darwin acknowledged that embryos could reveal structures of 'less modified ancient progenitors', just as extinct life forms might resemble the embryos of their descendants. 'Embryology rises greatly in interest,' he noted, 'when we thus look at the embryo as a picture, more or less obscured, of the common parent-form of each great class of animals.'[14] He also considered how such an evolutionary perspective helped to explain vestigial organs, which were otherwise explained in Cuvierite terms as created (as Darwin put it), 'for the sake of symmetry', or in order 'to complete the scheme [i.e. plan] of nature'. 'But this seems to me no explanation,' Darwin argued, 'merely a restatement of the fact.'[15]

'Nature may be said to have taken pains to reveal' by such vestigial organs 'her scheme of modification, which it seems that we willfully will not understand', Darwin noted in his conclusion. He noted how his theory would make all departments of 'natural history' more interesting as well as more revealing. Other fields such as psychology would find in it a new

---

[12]Ibid., p. 227.
[13]Ibid., p. 221.
[14]Ibid., p. 282.
[15]Ibid., p. 284.

basis for endeavour. 'Light will be thrown on the origin of man and his history,' Darwin promised. Given the complete failure of *The Origin* to consider man's ape origins, this was a curiously teasing remark.[16] Despite his reassuring claims to the contrary, his theory suggested a universe led by random mutations, by chance, rather than one sustained by God. Far from being a smiling, Paleyite one, for Darwin the ruthless battle for life made 'the face of Nature' appear like 'a yielding surface, with ten thousand sharp wedges packed close together and driven inwards by incessant blows, sometimes one wedge being struck, then another with greater force'.[17]

Darwin's bulldog

'What a book it is!' wrote the journalist Harriet Martineau on reading *The Origin*, 'overthrowing (if true) revealed Religion on the one hand, & Natural (as far as Final Causes & Design are concerned) on the other'. Martineau had been raised in a Unitarian family and hence espoused a version of Christianity which denied the divinity of Christ and proclaimed necessitarianism (i.e. the doctrine that all actions are predetermined, that there is no free will). Martineau lived off her pen, writing novels, Malthusian tales of political economy as well as accounts of her many travels. Her public profile and intellect struck many male as well as female contemporaries as 'unwomanly'. Given her faith and Malthusian views of the hard laws of political economy, it is perhaps unsurprising that she welcomed Darwin's book. If anything she felt he had pulled his punches in going 'out of his way two or three times . . . to speak of "the Creator" in the popular sense of the First Cause'.[18]

Others felt that Darwin had gone too far. These included Darwin's mentor, Henslow, who described *The Origin* as offering a hypothesis rather than a theory. Rather than simplifying their model of nature's machinery, both Henslow and Sedgwick felt that natural selection had complicated it. 'You have *deserted* . . . the true method of induction,' the latter wrote, 'and started in machinery as wild, I think, as Bishop [John] Wilkins's locomotive that was to sail with us to the moon,' referring to that seventeenth-century theologian's science fiction writing. Man would, Sedgwick insisted, be brutalized if his world was represented as one without divine love.[19] In public Sedgwick gave the book a hostile review in the *Spectator*. So did Owen. Owen, Darwin suspected, resented any challenge to his position as the unofficial doyen of British natural science. The harshness of Owen's attack in

---

[16]Ibid., p. 306.

[17]Ibid., p. 43.

[18]Harriet Martineau to Erasmus Darwin and Frances Wedgwood, 2 February and 3 March 1860. Cited in Desmond and Moore, *Darwin*, pp. 486–7.

[19]Adam Sedgwick to Charles Darwin, 24 November 1859. Darwin Correspondence Database, http://www.darwinproject.ac.uk 2548 (accessed 2 May 2013).

the April 1860 issue of *The Edinburgh Review* nonetheless shocked Darwin. But controversy sold books. As early as December 1859 John Murray had decided to publish a second edition, this time of 3000 copies.

On 26 December 1859 Huxley's anonymous review of *The Origin* appeared in *The Times*, a popular newspaper with a readership far larger than any monthly review, one which normally did not review works of natural history (it was, by contrast, normal for reviewers in this and many other papers to be anonymous). Huxley noted that natural historians had largely refused to consider how speciation occurred, in spite of the embarrassing species problem.

> Since Lamarck's time, almost all competent naturalists have left speculations on the origin of species to such dreamers as the author of the 'Vestiges', by whose well-intentioned efforts the Lamarckian theory received its final condemnation in the minds of all sound thinkers. Notwithstanding this silence, however, the transmutation theory, as it has been called, has been a 'skeleton in the closet' to many an honest zoologist and botanist who had a soul above the mere naming of dried plants and skins. Surely, has such an one thought, nature is a mighty and consistent whole, and the providential order established in the world of life must, if we could only see it rightly, be consistent with that dominant over the multiform shapes of brute matter. But what is the history of astronomy, of all the branches of physics, of chemistry, of medicine, but a narration of the steps by which the human mind has been compelled, often sorely against its will, to recognize the operation of secondary causes in events where ignorance beheld an immediate intervention of a higher power?

Though it would, Huxley suggested, take 20 years or more to investigate whether natural selection could in fact explain everything Darwin said it did, *The Origin*'s 'most ingenious hypothesis' provided a sound basis for scientific investigation.

The facts Darwin cited could be 'brought to the test of observation and experiment', unlike the 'cobwebs' necessarily woven by those who insisted on the immutability of species. The variability of life found in artificial selection and in the fossil record, as well as the presence of vestigial organs, all suggested that 'all living beings march, side by side, along the high road of development, and separate the later the more like they are'.[20] Huxley cheekily compared this to a congregation filing out of a church after the service: some people peeled off at the rectory, others at the end of the lane, while some kept company with each other beyond the parish boundary. It was not the last time Huxley used ecclesiastical similes in support of theories many churchmen found profoundly upsetting.

Writing to Huxley from a spa in July 1860 Darwin noted that *The Origin* would have been '*utterly* smashed had it not been for you &

---

[20]The Huxley File, http://aleph0.clarku.edu/huxley/CE2/Hypo.html (accessed 2 May 2013).

three others. I can now see (which I did not at first) how *very* [underlined three times] important the few early favourable reviews were'.[21] From being his 'bugbear' Huxley had become his 'bulldog'. Darwin's letter also acknowledges Huxley's support at that month's meeting of the British Association for the Advancement of Science (BAAS), held in the Library of the new University Museum of Natural History in Oxford (when Darwin had been, as usual, absent). Invited to respond to a long and dull paper on 'Civilization according to the Darwinian hypothesis' the Bishop of Oxford, Samuel Wilberforce had engaged in a critique of *The Origin*. The night before Owen had helped Wilberforce prepare his arguments. Drawing to a close, Wilberforce attempted to lighten proceedings with a topical joke, turning to ask Huxley if he was descended from an ape on his grandmother's or grandfather's side.

Huxley climbed onto the dais and delivered his reply:

> If then, said I, the question is put to me would I rather have a miserable ape for a grandfather or a man highly endowed by nature and possessed of great means & influence & yet who employs these faculties & that influence for the mere purpose of introducing ridicule into a grave scientific discussion I unhesitatingly affirm my preference for the ape.

At this point the packed Library erupted into laughter. Lady Brewster fainted. Others tried to speak. Fitzroy stood among the audience, holding a Bible aloft with both hands, imploring the audience to trust in God, not in man – in this case, his former *Beagle* companion, Darwin. In the twentieth century the episode became a lovingly polished artefact of the Victorian struggle between a dying Church and the unstoppable march of Science.

Scholars who have gone in search of contemporary eyewitness accounts paint a very different, confused picture. If this was an epic contest, it seems odd that the press almost entirely overlooked it at the time. Several accounts, including that of Hooker, suggest that Huxley could not be heard, possibly because he had been enraged to the point of incoherence. Wilberforce thought he had won and Huxley had lost. Son of the anti-slavery campaigner William Wilberforce, he was far from being an ignorant cipher. Though his review of *The Origin* was hostile, Wilberforce insisted that it was timid, if not plain wrong, to dismiss a theory just because one believed it 'to contradict what it appears to them is taught by Revelation'.[22] The misleading account of the episode partly reflects twentieth-century views of the history of science as a struggle against the obstructionism of the churches.

The conflicting accounts of the BAAS dispute suggest the mood of the educated public in the immediate wake of *The Origin*: one of muddled

---

[21]Darwin to Huxley, 3 July 1860. Darwin Correspondence Project. http://www.darwinproject.ac.uk 2854 (accessed 2 May 2013).

[22]*Quarterly Review* 108 (July 1860), p. 256.

curiosity, perhaps a hunger for something that might crystallize this talk of 'development' into a simple question, say, of whether humans were descended from an ape or not. As should be obvious, such reductions were controversial, but hardly 'Darwinian'. The theory does not propose that humans are descended from apes, but that apes *and* humans are descended from a common ancestor. As historians, however, such faultfinding is unhelpful. We need to study not only what this or that evolutionist said, but also (more importantly, perhaps) what their audience heard.

Darwin's illness and personality limited his ability to influence the direction of the debate started by *The Origin*. He did, however, take steps in 1861 to publicize a collection of articles defending *The Origin* that had been written for the American magazine *The Atlantic* by Asa Gray, professor of Natural History at Harvard University. A botanist, Darwin briefly met Gray in London in 1855 and was soon pestering him with questions about the distribution of plants in North America. As we have seen, Darwin sensed that Gray was unsympathetic to his colleague Agassiz's idealist approach, and drew him into the inner circle of those who knew the direction of Darwin's thought before 1859.

Gray fulfilled a somewhat similar role to Huxley when it came to *The Origin*. Although he was not convinced, he was determined that the theory would have fair play on his side of the Atlantic. He supervised the American edition on Darwin's behalf, when most British authors had to put up with seeing their works badly reproduced in American pirate editions, there being no arrangement to protect their copyright outside the United Kingdom. Gray's essays in *The Atlantic* argued that, far from being dangerously materialist, the theory of natural selection had reintroduced teleology into the natural world. It was a useful hypothesis, and, in the absence of a theory to explain where mutations came from, it was perfectly possible for one 'to assume, in the philosophy of his hypothesis, that variation has been led along certain beneficial lines'.[23] Although he would have disagreed with that proposal, Darwin clearly felt that Gray's response could serve a calming effect. After persuading a publisher to issue Gray's articles as *Natural Selection Not Inconsistent with Natural Theology* Darwin sent copies to Charles Kingsley, Samuel Wilberforce, Robert Chambers and others, and included an advertisement in editions of *The Origin*.

In Britain Darwin's theory became irrevocably associated with Huxley. 'I would as soon have died as tried to answer the Bishop in such an assembly,' Darwin wrote at the time.[24] The effects of this dependence were profound. Among other things, it helps explain why episodes like the 1860 BAAS

---

[23]Cited in A. Hunter Dupree, *Asa Gray. 1810–1888* (Cambridge: Belknap, 1959), p. 296.

[24]Darwin to Huxley, 5 July 1860. Darwin Correspondence Database, http://www.darwinproject.ac.uk 2861 (accessed 2 May 2013).

dispute would later be recorded as science slapping down the church and why Richard Owen's reputation stands so low. Although the introduction considered Huxley's intellectual framework in some depth, it is high time to introduce this self-described renegade properly.

The youngest of six children born to what a Victorian would have called 'distressed gentlefolk' (i.e. middle-class people who had fallen on hard times), Huxley had only 2 years of formal education, before being farmed out to various relations. He began medical training in London, gaining a scholarship to train at Charing Cross Hospital in 1842. When his scholarship ran out, Huxley joined the Navy as a surgeon, hoping to raise money to pay off debts, his family being unable to support him. Charged with surveying the Great Barrier Reef off the Australian coast, *HMS Rattlesnake* set off in December 1846. On board Huxley decided to focus on jellyfish and sea nettles. He also studied the mollusc archetype. Unlike Darwin, Huxley read German, and his study of such archetypes, as well as his later embryological work, was strongly influenced by German *Naturphilosophie* and the work of Karl Ernst von Baer and Ernst Haeckel. He found *Vestiges* unpersuasive.

On *HMS Rattlesnake*'s return in 1850 Huxley joined the various learnt societies and began presenting papers. Owen helped persuade the Navy to part-fund the scientific memoir of the voyage. Though Huxley was helped and flattered by such patronage, it was barely enough to live on, and certainly not enough for Huxley to be able to bring his fiancée, Henrietta Heathorn (whom he had met in Sydney, during the *Rattlesnake* voyage) back to London to settle. A year after his return he wrote to Henrietta of his miserable condition. Although he could have papers published in any of the learnt societies' journals he wrote, none of them had a large enough circulation to be able to pay contributors. 'A man who chooses a life of science chooses not a life of poverty – but as you can see a life of *nothing*,' he wrote. 'Why persevere in so hopeless a course? At present I cannot help myself.'[25]

It would be a 5-year engagement, ending only when Huxley was appointed to a professorship at the School of Mines in 1854. The following year he became Fullerian Professor of Physiology at the Royal Institution, also in London. He began delivering a series of lectures to genteel audiences. His rise thereafter was swift. But he never forgot the turmoil of the 1840s and the disappointments of the early 1850s, for which he blamed Owen, unfairly. In 1858 he was publicly ridiculing Owen's vertebrate theory of the skull. Three years later he started a particularly vicious feud with Owen over the question of whether humans were the only creatures to have a

---

[25]Thomas Henry Huxley to Henrietta Heathorn, 4 May 1851. Imperial College, London. Huxley Papers, Hen146. Partly reproduced in Leonard Huxley, *Life and Letters of Thomas Henry Huxley*, 2 vols (London: Macmillan, 1900), 1: 68.

hippocampus major (the name of one of the folds at the rear of humans' walnut-shaped brains).

In a Royal Institution lecture delivered in March 1861, Owen had asserted that only humans had this feature, hence putting a certain amount of clear water between *Homo sapiens* and the apes. Huxley roundly accused Owen of lying. The feud raged for several years, helped by Huxley's gift for set pieces. At the 1862 meeting of the BAAS in Cambridge Huxley was pleased to spot the conservator of the Hunterian Museum, William Flower, in the audience. Flower happened to have a monkey brain in his pocket, which he was only too happy to dissect on stage, extracting the hippocampus. Satirical pamphlets were also published, including one penned by Charles Kingsley.[26] Though Darwin dodged this question in *The Origin*, thanks to Huxley 'Darwinism' became irrevocably associated with ape ancestry in the educated public's imagination.

A fervent admirer of Goethe, from a young age Huxley saw himself as a long-haired, Romantic hero, a Faust-like figure wrestling alone with demonic discoveries. This blind faith in his own destiny made him a charismatic performer at the lectern. One minute he was the artist possessed, quickly drawing sketches of a skull on the blackboard; another a hell-fire preacher, pounding those stupid parsons who filled parishioners' heads with Paleyite pap; at other, more calmer moments he could be charmingly funny, even confiding, one arm draped around a gorilla skeleton, like a music-hall comedian. Apart from the odd photograph, today we have to rely on works like *Evidence as to Man's Place in Nature* (1863) to get a sense of his style.

Based on his lectures, *Man's Place* focused on proving man's ape ancestry and had as its frontispiece a lineup of various skeletons (see cover) which would become a classic evolutionary image (along with Darwin's tree). Although Huxley addressed technical aspects of bipedalism, embryological recapitulation and the thorny question of whether apes had a hippocampus, he used homely metaphors and a ready wit in doing so. He invited his audience to join him in that great process by which 'the human larva' was gradually freeing itself from its hard shell-like pupa case. Every citizen, he insisted, had a duty 'to ease the cracking integument to the best of his ability'.[27] The Darwinian hypothesis was, for the present at least, the best tool we had to do this. In presenting a picture of progress, Huxley argued, it provided the thoughtful evolutionist with faith that man might attain 'a nobler future'. The endowment of speech (Huxley did not indicate who or what 'gave' this trait to humans alone) gave us the means, allowing us to accumulate, share and organize all of human experience.[28]

---

[26]Charles Kingsley, *The Speech of Lord Dundreary in Section D . . . on the Great Hippocampus Question* (London: n.p., 1862).

[27]Alan P. Barr (ed.), *The Major Prose of Thomas Henry Huxley* (London: University of Georgia Press, 1997), p. 66.

[28]Ibid., pp. 107–8.

The Descent of Man

By the time of the 1866 meeting of the BAAS in Nottingham the *Guardian* newspaper could report that Darwin's ideas were 'everywhere in the ascendant'. That year the proceedings were led by William Robert Grove (1811–96), a Welsh lawyer whose experiments in electricity had led him to invent the nitric acid battery as well as to develop a principle of the 'correlation of forces'. Groves' 1846 book *On the Correlation of Forces* emphasized the conversion of one form of energy into another (heat into electric current, say), a key insight which would underpin the first Law of Thermodynamics, a fundamental law of physics. Like Huxley Groves had struggled to support himself financially and accused learnt societies of nepotism, claiming that he could only get his papers published in the Royal Society's journal *Philosophical Transactions* by first making friends with an influential Fellow of that Society. Grove related the slow, incremental process of Darwinian evolution with the slow, incremental growth of the British constitution. Nay, more than that, 'our language, our social institutions, our laws' were 'the product of slow adaptations, resulting from continuous struggles . . . we follow the law of nature and avoid cataclysms'.[29]

As usual, Darwin did not attend the BAAS, but remained at Down. His experiments with pigeons and plants bore fruit in the shape of a book, *The Variation of Animals and Plants under Domestication* (1868). Once again the initial print run of 1,500 sold quickly. The work provided copious evidence of the plasticity of species. More importantly, it also saw Darwin propose a mechanism by which traits were inherited: 'pangenesis'. Each of the body's organs sent 'gemmules' containing information on their formation and structure to the sexual organs, where they were fitted together into a complete instruction manual for forming a complete body. The theory met with a lukewarm reception from all but Wallace.

In 1871 Darwin published *The Descent of Man, and Selection in Relation to Sex*. Though his 1830s notebooks show an early interest in the development of human and animal instincts, in *The Origin* Darwin had left their relationship alone, merely noting that his theory could bring new light to the question of humanity's origins. Now he tackled 'the descent or origin of man' directly. Darwin's revisions to later editions of *The Origin* (such as the 4th edition, published in 1869) had seen him increase the role given to environmental factors in driving speciation. This was partly a response to joint criticisms by the electrical engineer Fleeming Jenkin and his friend, the physicist William Thomson. Jenkin pointed to problems explaining how beneficial mutations would not be lost through 'blending' inheritance. Thomson used Fourier's Law of Conduction to work out how much time would be needed for the earth to reach its current temperature, assuming a slowly cooling earth.

---

[29]W. R. Grove, *The Correlation of Physical Forces*, 5th edn (London: Longmans, 1867), p. 346.

Thomson's estimate cut Darwin's 300m years by a third, leaving too little time, as it were, for life to have developed by random mutations. In *The Descent* Darwin conceded that he might have overestimated the power of natural selection in *The Origin*, partly, he explained, for tactical reasons, that is, to get readers to abandon the idea of special creations. Lamarckian use/disuse was also evident in *The Descent*, a far longer and leggier work than *The Origin*. 'We may feel assured,' Darwin wrote, 'that the inherited effects of the long-continued use and disuse of parts will have done much in the same direction with natural selection.'[30]

In part one of *Descent* Darwin carefully removed the various barricades and barriers which had been thrown up by those who refused to accept that humans had descended from ape-like ancestors by a process of unguided natural selection. The Duke of Argyll, for example, accepted natural selection, but felt that mutations were guided, rather than random. Wallace agreed that natural selection explained the development of 'lower' creatures, but saw the last step to humans as too much of a leap to be explained by such natural forces. Hence his embrace of spiritualism, considered in the final chapter. Among the human endowments which Darwin sought to explain by natural selection and other unguided evolutionary mechanisms were tools, language, sympathy, altruism and other 'higher moral faculties'. He cited examples of animals behaving 'heroically' and explained how language could have evolved from simple animal cries.

Darwin considered how human sympathy could harm the species. By preserving the lives of the sick and maimed, for example, our acquired instinct to care might allow them to reproduce, causing the multiplication of the unfit, rather than their destruction in the struggle for life. He argued that we should not attempt to check our sympathy, as to do so would cause 'deterioration in the noblest part of our nature'. Without any revealed notion of what counted as 'noble', however, this argument was poor. Darwin changed the subject, only to return like a dog to a bone a few pages later. There he noted how the multiplication of 'unfit' (Roman Catholic) Irish immigrants was checked by the unsanitary conditions of the urban slums in which they lived, which killed many of their offspring. Although such views reflected English prejudices of the time, they also indicated a readiness to abandon 'nobler sentiments' in some cases.[31]

Darwin's insistence that 'the several so-called races of men' were one species (i.e. his support of monogenism over polygenism) seems reassuring, reflecting as it does his long-held abhorrence of slavery. Yet here again Darwin was confident that his people, the 'civilised' ones, would 'exterminate' the 'savage' peoples – and was totally unconcerned.[32] In part

[30]Charles Darwin, *The Descent of Man and Selection in Relation to Sex* (London: John Murray, 1922), p. 928.
[31]Ibid., pp. 206, 213.
[32]Ibid., p. 257.

two Darwin turned to consider sexual selection, noting (as in *The Origin*) that the average reader would find the powers of selection it gave to women 'extremely improbable'. In this rambling section Darwin considered sexual dimorphism from insects and snails through birds and mammals, turning to humans in part three.

Although he noted that in humans 'pugnacity' was a male trait, Darwin nonetheless insisted that the 'rapid perception' and greater intuitive powers he found in female humans were 'characteristic of the lower races, and therefore of a past and lower state of civilisation'.[33] This was illogical (elsewhere Darwin showed that he shared the consensus view that civilizations became less 'pugnacious' over time) and presumably showed Darwin reflecting the prejudices of his time.

Darwin turned 62 in 1871. Thanks to the many translations of his works, he enjoyed an international fame and collected honours from across the globe. When *The Origin* first appeared Marx and Engels had joked that it was a case of Darwin transposing the ruthless struggle of *laissez-faire* capitalism to the natural world. Marx sent him a copy of the second, 1873 edition of *Das Kapital*, writing on the flyleaf that it came from a 'sincere admirer'. Sadly, Darwin's German just wasn't up to reading more than a few pages. Darwin's last work to reach a sizeable audience was his *Expression of the Emotions in Man and Animals*, published in 1872 and illustrated with several photographic plates (still an expensive novelty). His final book on worms (1881) sold well, but one wonders how many of the copies purchased were actually read.

Charles Darwin died at Down, surrounded by his family, on 19 April 1882. After a public campaign led by Huxley and others, he was buried, not in the churchyard at Down, but in Westminster Abbey. Though the family acquiesced, one senses that the villagers of Down were right when to insist that their neighbour would have preferred to rest in Down. The village joiner, John Lewis, had made up an oak coffin, 'just the way he wanted it, all rough, just as it left the bench, no polish, no nothin'. But 'they' sent it back, using a machine-polished coffin 'you could see to shave in'.[34]

A Darwinian revolution?

In 1880 Huxley gave a lecture at London's Royal Institution entitled 'The Coming of Age of the Origin of Species,' celebrating the 21st anniversary of Darwin's landmark book. In the years since Darwin had gone into print, Huxley noted, there had been one of a number of discoveries (such as the lizard bird *Archaeopteryx*) which had come to populate those neat, white

[33]Ibid., p. 858.

[34]R. Colp, 'Charles Darwin's Coffin, and its Maker,' *Journal of the History of Medicine and Allied Sciences* 35 (1980): 59–63.

margins which had previously separated birds from reptiles and vertebrates from invertebrates. Modestly, Huxley refused to be seen as Darwin's bulldog, adopting a less strident metaphor. 'I acted for some time in the capacity of a sort of under-nurse,' he recalled. As he noted, before 1859 everyone had held a catastrophist view of the earth's history. After 1859, everyone was uniformitarian, recognizing that 'the explanation of the past is to be sought in the study of the present'.[35] Darwin, in short, had triumphed.

Yet there are serious problems with this picture of an 1859 'Darwinian revolution'. Huxley's account of a pre-1859 consensus is incorrect: catastrophist views of geology and zoology were not characteristic. More importantly, the enthusiastic reception of *The Origin* did not mean enthusiastic reception (or even comprehension) of the ideas within it. As Wallace noted at the time (and as Darwin came to recognize), 'natural selection' as well as references in *The Origin* to it displaying anthropomorphic characteristics ('unerring skill', 'vigilance' and so on) could be taken as implying a personified 'selector' acting consciously. Darwin told Lyell that he would have preferred 'natural preservation'.[36] Such personality traits could be taken as implying a (divine) lawgiver, fostering a theistic naturalism that the Rev Charles Kingsley and other Christian evolutionists embraced wholeheartedly.

The most persistent misreading, commonly found today, was that which revived old 'chain of being' thinking. Victory in the 'struggle for life' was linked with progress, with an overarching teleology which moved from 'lower' to 'higher' life forms, with the human as the end point (and, within *Homo sapiens*, the British as the highest form of humanity). Darwin struggled in vain to get home the message that natural selection selected those traits which were 'fittest' only for the specific time and place in which the form then found itself. As we shall see, Britons proud of their Industrial Revolution, expanding empire and growing reputation for scientific discovery found this misreading of Darwin highly appealing. It went well with Herbert Spencer's fashionable theories of meliorist evolution as a 'beneficent necessity'. It went very well with Broad Church Anglicanism, as its adoption by the Rev Charles Kingsley indicated.

Darwin even failed to find many converts for natural selection among his close allies, or among others who shared his expertise in geology and zoology. Having chosen the tree metaphor to get away from anthropocentric 'chain of being' models, Darwin was powerless to see colleagues make a ladder out of it, by giving it a central 'main line' or trunk, which lead, arrow-like, straight to humans.[37] Even his co-discoverer, Wallace, had come

---

[35]Huxley, 'The Coming of Age of "The Origin of Species",' in Barr (ed.), *Major Prose of Huxley*, pp. 195, 198.

[36]Robert M. Young, *Darwin's Metaphor: Nature's Place in Victorian Culture* (Cambridge: Cambridge University Press, 1985), p. 95.

[37]Michael Ruse, *Life's Splendid Drama* (Chicago: University of Chicago Press, 1996), p. 425.

to bolt a guided, spiritualist teleology on top of natural selection, arguing that natural selection could not explain the leap from ape to human. After reading one paper Darwin wrote to Wallace, expressing the hope that Wallace had not 'murdered' what he called 'their baby'. In 1889, after Darwin's death, Wallace published *Darwinism: an exposition of the theory of natural selection with some of its applications*. In its closing pages Wallace argued that the moral, artistic and other 'higher' human faculties could only be explained by reference to 'an unseen universe – to a world of spirit, to which the world of matter is totally subordinate'.[38] This, 'Darwinism'?

Darwin's bulldog failed him, too. If we re-read 'On the Coming of Age' carefully we find no mention of 'natural selection' at all, only a haunting, perhaps even sinister observation or prediction that 'it is the customary fate of new truths to begin as heresies and to end as superstitions'. Huxley only seems to have been brought to abandon his belief in Owenite archetypes by his reading of German embryologists like Haeckel, whose *Generelle Morphologie* Huxley read in 1866. Before *The Origin* he could write that 'the doctrine that every natural group is organized after a definite archetype' as being 'as important for zoology as the theory of definite proportions for chemistry'.[39]

Huxley also clung to notions of an *Urschleim* or 'protoplasm', the basic goo of life, rather than adopting the cell theory (i.e. the theory which has the cell as the fundamental building block of all living things). When the research vessel *HMS Challenger* returned in 1868 with a curious goo in one of its jars Huxley rushed into print to announce the discovery of 'the physical basis of life' (even giving it a binomial classification: *Bathybius haeckelii*). This goo turned out to be a precipitate (i.e. random gunk), and Huxley had to concede an embarrassing error, more glaring than anything Owen might or might not have done. Even after his belated conversion he continued to avoid transmutation in his teaching, devoting less than half a lecture (out of 165) of the 2-year biology course he taught. As Michael Ruse has noted, 'Huxley could have been a Six Day Creationist for all that evolution counted in his professional activities.'[40]

Darwin's biographers put it pithily. Darwin 'had turned the world to evolution, and practically no one to natural selection, not even his champions'.[41] It is unhelpful, therefore, to speak of a 'Darwinian revolution', to follow Huxley in seeing 1859 as the 'year zero', the beginning of a new age of evolution. In its chronological structure and its focus on Darwin,

---

[38]Alfred Russel Wallace, *Darwinism* (London: Macmillan, 1889), pp. 473, 476.
[39]Michael Foster and Ray Lankester (eds), *The Scientific Memoirs of Thomas Henry Huxley*, 3 vols (London: Macmillan, 1898–1903), 1: 192.
[40]Michael Ruse, 'Thomas Henry Huxley and the Status of Evolution as Science,' in Alan P. Barr (ed.), *Thomas Henry Huxley's Place in Science and Letters: Centenary Essays* (London: University of Georgia Press, 1997), p. 147.
[41]Desmond and Moore, *Darwin*, p. 642.

the last four chapters may have been guilty of this, at least to some extent. Part two considers the different evolutionary models championed by Victorians. Rather than a single narrative strand it presents five different 'lines of descent'. Darwin takes something of a back seat as we consider four evangelists, as it were, of Darwin, men who, like Christ's evangelists, took the message and brought it to the people, changing it as they did so: Huxley, Kingsley, Spencer and Wallace. We also consider the different physical spaces in which Victorians read, spoke, argued and consumed evolutionary ideas, from the well-upholstered middle-class home to the gin palaces of gritty south London.

Further reading

Bowler, Peter, *The Non-Darwinian Revolution: Reinterpreting a Historical Myth* (Baltimore: Johns Hopkins University Press, 1988).

Browne, Janet, *Darwin's Origin of Species. A Biography* (London: Atlantic, 2006).

Desmond, Adrian, *Archetypes and Ancestors: Palaeontology in Victorian London, 1850–1875* (London: Blond and Briggs, 1982).

—, *Huxley. The Devil's Disciple* (London: Michael Joseph, 1994).

Richards, Evelleen, 'Darwin and the Descent of Woman,' in David Oldroyd and Ian Langham (eds), *The Wider Domain of Evolutionary Thought* (London: Reidel, 1983), pp. 57–112.

Ruse, Michael and Robert J. Richards (eds), *The Cambridge Companion to 'The Origin of Species'* (Cambridge: Cambridge University Press, 2009).

White, Paul, *Thomas Huxley. Making the 'Man of Science'* (Cambridge: Cambridge University Press, 2003).

PART TWO

Lines of Descent, 1850–1914

CHAPTER FIVE

Christian evolution? Charles Kingsley's 'natural theology of the future'

FIGURE 3 *John and Charles Watkins,* The Rev Charles Kingsley, *1860s. One of a vast number of 'carte de visite' photographs of celebrities printed by the London Stereoscopic and Photographic Company and avidly collected by the public in the 1860s, such images were also high-tech, the albumen print technique having been invented just a decade before.*

In July 1862 the Rev Charles Kingsley wrote to Huxley describing how profoundly he had been affected by Charles Darwin's *Origin of Species*. 'I am as one overwhelmed and astounded by the grand views Mr Darwin's theories open to me at every turn,' he wrote. 'I believe that he has inaugurated a new era to me, as well as to your strictly scientific men; for all natural theology must be rewritten during the next century, by the light of his hints—for they are no more than hints—but hints wh[ich] will be, when modified by fresh knowledge, the parents of a whole new science.'[1] Kingsley had written in similar terms to Darwin himself upon the book's publication, noting that, far from undermining faith in God, the theory of natural selection would lend additional grandeur to our conception of Him. Darwin quoted Kingsley's letter (anonymously) in the second edition, in order to show that his hypothesis did not represent a high road to atheism.

Kingsley is best known today as the author of *The Water-Babies* (1863), an evolutionary fairy tale about a chimneysweep named Tom. The book became a children's classic in the twentieth century, as did Kingsley's historical novels, such as *Westward Ho!* (1855), a swashbuckling sea adventure set in the sixteenth century, around the time that the Spanish Armada threatened to invade Britain. Kingsley is also known as the father of that variety of Anglicanism known as Muscular Christianity or Christian Manliness (Kingsley's preferred term). Christian Manliness celebrated cleanliness, physical strength and a courageous enthusiasm to stand up against bullies and 'fight the good fight'. Neither the children's author nor the champion of public-school ethics seems to belong among the ranks of what Kingsley called 'your strictly scientific men'. Nor does the vicar of the sleepy agricultural parish of Eversley, Hampshire. As a priest, Kingsley seems to be particularly out of place. Evolutionary science and religious faith are widely held to be locked in a struggle to the death.

Familiar evolutionary set pieces like the 1860 British Association for the Advancement of Science (BAAS) 'debate' between Huxley and Bishop Wilberforce can encourage this sort of thinking. As we saw in the previous chapter, it is surprisingly difficult for today's historians to discover exactly what Huxley said on that famous occasion. Witnesses do not seem to have viewed it as an epic science-versus-faith struggle at the time. Yet the lines of a familiar battle were already being drawn. The Huxley/Wilberforce exchange took place immediately after a paper by the American John Draper, entitled 'On The Intellectual Development of Europe', which described the march of secularism in Darwinian terms. Some Victorian Christians certainly felt that their faith was being menaced by the geologists, who seemed to be hammering away at Holy Scripture, that impregnable (supposedly) rock. The great social critic and writer on art John Ruskin dreaded their hammers. 'I hear the clink of them at the end of every . . . Bible verse,' he anguished.[2]

---

[1]Kingsley to Huxley, 18 July 1862. Imperial College, London. Huxley Papers. Gen. Letters IX, f. 205.

[2]Tim Hilton, *John Ruskin: The Early Years* (New Haven: Yale, 1985), p. 167.

Kingsley refused to let the hammers scare him. He engaged with Lyell, Darwin, Huxley, Wallace, Asa Gray and many other 'strictly scientific men', both in print and in private correspondence, refusing to leave the field of natural history to what we, from our twenty-first-century perspective, would call 'the professionals'. And they took notice. As we have seen, Darwin incorporated Kingsley's response in *The Origin*. Huxley favourably reviewed Kingsley's guide to seaside natural history in the *Westminster Review*.[3] Such attention recognized Kingsley's effectiveness as a communicator. Behind Huxley's bluster about 'smiting' the Anglican clerics until they fled from the natural sciences, he and other evolutionists recognized that Kingsley could be useful to them. For our purposes Kingsley is significant as an eminent Victorian who did not see evolution and faith as in conflict. Kingsley took significant numbers of fellow Protestants (in particular that group of Anglicans known as Broad Church Anglicans) with him as he sought a constructive relationship between faith and 'the new science'.

Kingsley had struggled with the Oxford Movement in the 1830s and with Anglican orthodoxy in the 1840s, thanks to his role as spokesman for Christian Socialism and his doubts about the existence of hell and the devil. In the 1850s he became concerned at the widening gap between theology and the natural sciences, blaming church dogmatists as much as men of science for creating it. Kingsley nonetheless found in Darwin the inspiration for a 'natural theology of the future', one which would, eventually, reunite theology and the natural sciences. He needed Huxley and Darwin as much as they needed him. Just because Kingsley and the 'strictly scientific men' needed each other does not mean that they were allies, however.

This chapter begins with a brief account of Kingsley's career and works, before looking in depth at how he used ideas of superfecundity, competition, degeneration and recapitulation. It then considers Kingsley's concern that men of science were becoming seduced by their own evolutionary 'laws', giving those laws greater status and authority than they deserved. Such scientism (excessive belief in the power of scientific knowledge and techniques) harmed both theology and natural science. Though Kingsley died in 1875, as we shall see he was not alone in this fear and in his desire to place the 'secondary laws' of natural science within a guided or otherwise Providentialist universal order.

The apostle of the flesh

Kingsley was born in Devon in 1819, the third son of an Anglican priest. Kingsley's father held a series of livings in Devon during Charles' youth. The family had been wealthy enough for Kingsley's father to have attended Harrow and Oxford, but Charles was sent to less exclusive preparatory

[3]Huxley, 'Science,' *Westminster Review* 8 (1855): 246–53.

and grammar schools in Clifton, near Bristol, and Helston, in Cornwall. His mother's people had been wealthier, plantation owners on Barbados. Her low-church Evangelicalism set the tone for the household. Part of that late eighteenth-century spiritual revival of the Church of England associated with John Wesley, Evangelicals tended to emphasize the authority of the Bible over that of the church. They drew a clear divide between the 'saved', those who had 'come to Jesus' in a sharp, sudden, personal conversion experience, and the rest, the 'damned'. The world was a sinking ship tainted by sin, but the Second Coming of Jesus was nearly here, the time when Jesus would separate the sheep from the goats, casting the 'damned' into eternal fire and inviting the 'saved' to join him in Heaven where eternity was to be spent praising God.

Kingsley seems to have found the resulting mood overly strident and the flurry of genteel parish rituals stifling. At school his bookish habits and a stammer isolated him somewhat. In 1831, while at Clifton, Kingsley snuck away to witness the immediate aftermath of a pro-Reform riot. Three years later, at Helston, his younger brother Herbert ran away from school, only to be captured and returned. The shame of being found in possession of stolen property may have contributed to Herbert's subsequent death by drowning, a tragedy Charles may have understood as suicide, rather than as accident. Inexplicable, inscrutable, both events haunted Charles. On going up to Cambridge in 1838 Kingsley threw himself into the hard-rowing, heavy-drinking life of a 'hearty', cramming for exams and attending Sedgwick's geological lectures and field trips intermittently. On 4 July 1839 he met Frances (Fanny) Grenfell, one of four sisters born to a well-off industrialist, and everything changed. Like a hero from a medieval romance, he vowed to put himself to the test, to prove himself worthy of marrying her.

In the years around 1840 the model followed by the Darwins was still the norm when it came to marriage: hard-headed profit-and-loss accounting (weighing up the partner's wealth, social class, age, prospects of professional advancement) on the part of the families concerned, accompanied by the writing of romantic letters and an eventual proposal, the whole exercise providing the servants with a few weeks' good gossip. Expectations were low, and frequently met. If a loving, companionate union emerged, it was a pleasant surprise, as one senses it was for Charles and Emma Darwin. Kingsley's courting of Fanny and their marriage did not fit this mould. Fanny and her sisters were followers of Edward Pusey, one of the leaders of the Oxford Movement, in other words High Church, the opposite end of the Anglican spectrum from Kingsley's mother. In style of worship and in doctrine Evangelical or Low Church Anglicans resembled Methodists more than they did High Church Anglicans, who themselves resembled Roman Catholics. In 1844 one High Churchman, William G. Ward published a work entitled *The Ideal of the Christian Church* that urged the Anglican church to follow the model of the Roman Catholic church. Determined to contain the Oxford Movement, Oxford University stripped him of his degree.

Pusey advocated clerical celibacy (as the Roman Catholic church still does) and would later (1846) advocate individual confession of sins, made privately to a priestly confessor (again, the Roman Catholic way; Anglicans make a public confession, as a congregation). Fanny was resolved to join her sisters in a celibate Puseyite sisterhood, Park Village. To persuade her otherwise Kingsley had himself ordained a deacon, then a priest, based at Eversley, Hampshire. The pair engaged in a long and passionate courtship by letter, debating the merits and demerits of sex and abstinence. Struggling with themselves and with each other, the correspondence and the erotically charged drawings Kingsley produced continue to fascinate and titillate us today. The self-flagellation (real and metaphorical) involved can be seen as deviant psychological tics unique to Kingsley. As historians, it is more helpful to note how closely they reflect the charged mood of the years around the publication of *Vestiges*, Newman's 1845 conversion to the Roman Catholic church and the debate over state funding of the Roman Catholic seminary at Maynooth in Ireland. For thoughtful members of Kingsley's Oxbridge cohort in the 1840s the search for a fulfilling goal or vocation in life was far more fraught than it had been for Darwin's 10 years before.

The British state continued cutting its ties to the Established church, to the Church of England. Admitting Jews to parliament and letting non-Anglicans off paying tithes to their Anglican parish priest seemed liberal, tolerant, inevitable to Victorians like the historians T. B. Macaulay and Henry Buckle, part of a universal march of 'mind' or 'civilisation'. To others, including Kingsley, as well as the future Liberal prime minister William Gladstone, this toleration felt like indifferentism. It implied that religion simply involved a private choice among equally valid alternative churches, a choice without any repercussions for public or community life, which thereby lost any sense of spiritual identity or collective 'personality'. Being left alone to practice your own religion was not enough, because the values that religion taught were also those required to hold society together.

Could such an eviscerated, hollowed-out state continue to inspire its citizens to put aside self for the greater good? Or would it leave them as isolated atoms, as profit-and-loss utilitarians, worrying about what services the state provided them 'in return for' so-and-so-much payment in the form of taxes? Would it do the working man any good to give him the right to elect representatives within such a state, as the Chartists were demanding? Was parliament simply the sum of all these representatives' voices? 'Of course it is!' we reply today. Inspired by the poet Samuel Taylor Coleridge, many Victorians held that parliament could be more than this sum, that it had a collective mind or conscience (something innately spiritual, dressed in religious forms and rituals) that could lead the people, all of them, rather than simply following the majority of those who happened to have a vote in this or that election. Thoughtful Anglicans like Baden Powell and the young Gladstone gradually came to the conclusion that the best option was Disestablishment (the cutting of all ties between the state and the Church of

England). The decision that those who loved the Church needed to set it free was a drawn out and agonizing one.[4]

Charles' desire to throw himself into this church and state ferment did not vanish, therefore, when Fanny and Charles were wed in January 1844. Kingsley wrote *The Saint's Tragedy*, a play about the thirteenth-century Saint Elizabeth of Hungary, which provided its author with a suitable vehicle for criticism of clerical celibacy. In it the Teuton character, Walter of Avila, urges the King of Hungary to rule manfully and to make a woman of his betrothed Elizabeth. He is contrasted with Conrad, Elizabeth's confessor, described as typical of those 'sleek passionless men, who are too refined to be manly, and measure their grace by their effeminacy'.[5] In 1844 Kingsley met Frederick Denison Maurice, the theologian behind Christian Socialism. Whereas Evangelicals prayed for Jesus to come and reign, Maurice insisted that His Kingdom was already with us, here on earth. His image could be recognized in flawed fellow mortals and in nature. Rather than attempting to remain aloof from a 'sinful' world, Christians' duty was to get to grips with the world as it was and seek to improve it.

Kingsley acknowledged Maurice as 'Master'. Eager to escape the quiet of Eversley, in 1848 Kingsley eagerly accepted Maurice's invitation to lecture on English Literature at Queen's College, a girls' school Maurice had established in Harley Street. Meanwhile a novel, *Yeast*, began coming out in instalments in *Fraser's Magazine*. Kingsley was attempting to tap into the revolutionary mood of the times, using his characters as ventriloquists' dummies, voicing the dissatisfaction of his generation with the Church's doctrines, as well as their suspicions that a sacred purpose could be found outside their confines. *Yeast* was one of a series of novels published around this time that recounted young men's crises of faith, including Newman's *Loss and Gain* (1848) and J. A. Froude's *Nemesis of Faith* (1849).

The Chartist movement added to Kingsley's sense of frustration, of missing an appointment with destiny and with the seething, searching urban working classes, whom Kingsley only knew from the pages of Henry Mayhew's *London Labour and the London Poor* (1851). Drawn up in 1838, the Charter demanded not only universal manhood suffrage (i.e. votes for all adult men), but also a secret ballot (all votes were then cast in public), annual parliamentary elections and salaries for MPs (so that working men could support themselves if elected). It provided a diffuse, temporary alliance of trade union, socialist, religious and temperance (anti-drink) groups with a single, much-needed focus. Whether rushing to post up broadsides addressed to Chartists, writing 'communist' squibs attacking sweat shops (notably the 1851 Christian Socialist tract entitled *Cheap Clothes and Nasty*) or delivering controversial sermons urging the faithful to doubt their own parish priests,

---

[4]W. E. Gladstone, *Substance of a Speech on the Motion of Lord John Russell* (London: John Murray, 1848).

[5]Charles Kingsley, *The Saint's Tragedy* (London: n.p., 1848), p. xix.

there was a strong sense that Kingsley was trying to jump the bandwagon of working-class agitation in order to slow it down. Kingsley publicly declared himself a Chartist; in his second novel, *Alton Locke* (1850), he wrote as one, writing the autobiography of a fictional Chartist tailor and poet. Alton's fellow Chartists are depicted as idolizing the Charter as a panacea, as an impossible, once-and-for-all solution to all their problems.

Over the course of the novel Alton learns that salvation comes, not from a list of paper rights extracted from others by mass protest, but by individual, practical action to address poor sanitation, housing and intolerable working conditions. Social change is understood to be impossible outside a Christian framework; the Charter cannot take the place of Christ. This learning process is given an evolutionary gloss in the 'Dreamland' chapter, in which a delirious Alton imagines himself metamorphisized into a madrepore (a kind of coral) only to evolve 'back up' to *Homo sapiens*. Imprisoned for his part in a Chartist protest, Alton watches from his cell window as a parish priest sets to work improving his parish. Together with his Christian Socialist friends Kingsley gave a real-life demonstration by buying a water cart to deliver fresh water to Jacob's Island, a cholera-infested slum in south London, later made notorious by its depiction in Charles Dickens' novel *Little Dorrit* (1855–57). This emphasis on cleanliness and on evolution as a two-way street would feature strongly in *The Water-Babies*.

THE FAIRYLAND OF SCIENCE: *THE WATER-BABIES*

On one level, *The Water-Babies* is a story of sin, repentance and salvation, with a young chimneysweep, Tom, as its hero. In the beginning Tom is described as ignorant of right and wrong, sin and Christ, mutely accepting the blows given to him by his master, Mr Grimes, apparently unable to imagine how life could be any different. When Grimes is summoned to a rambling country house, Tom nimbly climbs up the chimney flues and sets about sweeping them. He gets lost and drops into the bedroom of little Ellie. She is there, innocently sleeping, a clean, white girl. Tom is transfixed, then sees himself, black with soot, in her mirror. For the first time he is aware of his sinful, 'dirty' state, as well as of his potential to be like Ellie, to be clean. Ellies awakes, sees Tom and screams. Tom runs away and staggers, dazed, into a village pond, muttering to himself 'I must be clean, I must be clean.' Suddenly he is underwater, transformed into a water-baby.

Tom follows the stream down to the sea, meeting Mrs Bedonebyasyoudid, who proceeds to teach him simple moral lessons using the sea creatures around them, including Ellie, now a fellow water-baby. Tom delights in her company, and is frustrated that he is left behind when she heads off to a special place every Sunday. He is told that to earn the right to do so he must 'do the thing he does not like', that is undertake a long, arduous journey to the Shining Wall and back. At the Wall he meets Mother Carey who creates all living things. He is disappointed to find her sitting calmly, when he expected to find her hard at work making this or that creature. 'But I am not going to trouble myself to make things, my little dear,' she explains. 'I sit here and make them make themselves.'[6] Tom returns and is rewarded with Ellie and with promotion back up the evolutionary ladder.

Kingsley's Creator is one who creates, but indirectly, by secondary laws. She is also a Judge. Unlike the Christian God, who will judge the living and dead after the Second Coming, this Judge is constantly making moral judgements of her creatures and rewarding/punishing them accordingly. Mrs Bedonebyasyoudid does not want her water-babies to see evolution as a one-way street:

> let them recollect this, that there are two sides to every question, and a downhill as well as an uphill road; and, if I can turn beasts into men, I can, by the same laws of circumstance, and selection, and competition, turn men into beasts. You were very near being turned into a beast once or twice, little Tom. Indeed, if you had not made up your mind to go on this journey, and see the world, like an Englishman, I am not sure but that you would have ended as an eft [a newt] in a pond.[7]

When Tom behaves badly he degenerates, developing nasty, ugly prickles, eating horrible things.

The Water-Babies can be a confusing read. The characters within Kingsley's fable tell fables themselves, some of them set in the same watery fairyland, others in jungles, yet more in what would then have been recognizably everyday settings. The authority figures (including the narrator) one normally looks to in fairy stories to provide clarity repeatedly refuse to do so. Thus the narrator tells his readers that they are not to consider his story as anything more than a fairy tale, 'even if it is all true'. This isn't a case of Kingsley writing sloppily or lazily. It reflects his hostility to the rote learning common in Victorian schools. He wants readers to find things out for themselves, not sit passively while their brains are programmed with knowledge they do not understand.

The Bermondsey water cart may have been a short-lived stunt (the locals stole the taps, then the barrels) and *Alton Locke*'s ending unsatisfying, but there was no doubting the conviction and energy behind both. In just 3 years (1848–51) Kingsley had become the most conspicuous and controversial spokesman of Christian Socialism, deliberately provoking elite public opinion in order to raise the movement's profile, and his own. As the 1850s progressed he began moving towards the centre ground, however, shifting with the times. The Chartist movement fizzled out after a mass demonstration on Kennington Common (in London) in 1848. The spectacle of the 1851 Great Exhibition in Hyde Park heralded an age of peaceful international

---

[6]Charles Kingsley, *The Water-Babies. A Fairy Tale for a Land-Baby* (London: Macmillan, 1895), p. 165.
[7]Ibid., p. 144.

trade, technological innovation and improvements to the welfare of the working man, a so-called 'age of equipoise', to contrast with the economic downturn and working-class suffering of 'the hungry forties'. Christian Socialism survived, but largely in the form of the cooperative movement, which Kingsley was uninterested in. Whereas publishers had quailed over his provocative first two novels, Queen Victoria and her consort Albert were delighted by his third, *Hypatia* (1853), a historical novel set in fifth-century Alexandria.

Although Sir Walter Scott had made the genre fantastically popular in the years before the Great Reform Act, by 1850 Scott's novels were felt to be dull. Kingsley's historical novels revived the genre, and, though neglected today, have survived better than those of his closest rival, Edward Bulwer-Lytton. Timed to coincide with the Crimean War, *Westward Ho!* (1855) was a patriotic journey back in time to 1588, when Sir Francis Drake and his brave sailors defeated the Spanish Armada, protecting Queen Elizabeth's Protestant, prosperous 'Merry England' from invasion by a Roman Catholic power. The novel's hero is a young Bideford roustabout named Amyas Leigh, eager to see new shores, perform brave deeds and return to claim local beauty Rose Salterne. His adventures first take him to dark, backward Ireland, then on to the Caribbean and South America, whose natives groan under the persecution of the Spanish and their perfidious Inquisition.

From the opening scene, in which Amyas breaks his slate over his teacher's head, through a massacre in Ireland and on the novel is unsentimental in its approach to violence, excusing it as high-spiritedness. Elizabethan England is a time of intellectual freedom, but also 'of immense animal good spirits', when the English conquered a new empire 'with the laughing recklessness of boys at play'.[8] Kingsley's willingness to excuse his boys any act of violence is evident in many of his works, particularly his last historical novel, *Hereward the Wake* (1866), in which the hero is 'the last of the race' of independent, feisty Saxons, fighting a one-man guerrilla war against the Normans in the years after the Norman Conquest (1066). Like the Spanish in *Westward Ho!*, so in *Hereward the Wake* the Norman enemy is portrayed as effeminate and priest-ridden.

Thanks to novels such as these Kingsley had gained enough of a reputation as a historian for him to be appointed, with Prince Albert's help, to the Regius Professorship of History at Cambridge University in 1860. This gave Kingsley the opportunity to develop his racial model of English history, a story which began when the Teutons (also known as Goths, or Huns) defeated the Romans in 9 CE and came south to conquer Rome and absorb Roman learning, before overrunning the rest of Europe. Kingsley could not agree with Macaulay that 'nothing in the early existence of Britain indicated the greatness she was destined to attain': race was destiny, and the Teutons were favoured by God with an intelligence, strength and resourcefulness

[8]Charles Kingsley, *Westward Ho!* (London: Collins, 1910), p. 43.

denied to other races.[9] As the heirs of the Teutons and leaders of an empire larger than anything the Romans could have dreamed of, it was the Britons' sacred duty to subdue the world. The Teutons had kept on going when other races had stopped, exhausted. They were, in short, history's last reserve.

At Cambridge Kingsley moved his students in ways no other Regius Professor of History had. Undergraduates cheered until Kingsley stammered for them to stop. A larger lecture hall had to be found. Then a larger one. 'Wild young fellows' eyes would be full of manly, noble tears,' one student recalled, 'And again and again, as the audience dispersed, a hearer has said, "Kingsley is right – I'm wrong – my life is a cowardly life – I'll turn over a new leaf, so help me God."'[10] But Kingsley also took advantage of the opportunity his position gave him to make Darwin's views better understood in his former university, as Darwin gratefully acknowledged. 'It is very interesting and surprising to me that you find at Cambridge after so short an interval a greater willingness to accept the views which we both admit,' Darwin wrote to Kingsley in late 1867. 'I do not doubt that this is largely owing to a man so eminent as yourself venturing to speak out.'[11]

As far as the Anglican church hierarchy was concerned Kingsley never entirely shook off the radical associations of the 1840s. At Oxford Pusey successfully blocked plans to award Kingsley an honorary degree. With Queen Victoria's help, however, Kingsley managed to secure nomination to fairly well-endowed positions in the 1860s and 1870s, at Chester Cathedral and eventually at Westminster Abbey. Fanny's taste for home improvements and needy relations took their toll on his finances, however. Encouraged by his publisher, Alexander Macmillan, Kingsley arguably wrote more than he should have, faster than he should have. In 1874 he travelled to America, having fallen victim to that golden mirage, the US lecture circuit. There he flattered his audiences as latter-day Teutons, back in the forests (of North America, this time, not North Germany), carrying the race ever westwards. But he failed to make the fast buck. Kingsley returned, exhausted and ill, and died the following year. Offered Westminster Abbey (on somewhat better grounds than Darwin), Fanny insisted that he be buried modestly in his home parish of Eversley.

Reproduce, rinse, repeat

The Water-Babies reflects the importance Kingsley places on imagination and fantasy in the pursuit of 'science', understood in its original sense,

---

[9]Thomas Babington Macaulay, *History of England* in *Works*, 12 vols (London: Longmans, Green, 1898), 1: 4.

[10]Fanny Kingsley, *Charles Kingsley: His Letters and Memories of His Life*, 9th edn, 2 vols (London: Macmillan, 1881), 2: 118–19.

[11]Charles Darwin to Charles Kingsley, 13 December 1867. Darwin Correspondence Database, http://www.darwinproject.ac.uk 5728F (accessed 2 May 2013).

as 'knowledge'. 'Have you lived before?' asks the narrator at one point, inviting his readers to reflect on 'their' earlier lives as apes or fishes, to use evolutionary thought to question who exactly 'you' think 'you' are. Am 'I' my personality, part of my species, an evolutionary lineage? Such 'make-believe' is not the enemy of truth and discovery. On the contrary, Kingsley insists, our ability to discover new truths, new knowledge is directly related to our ability to imagine, to fantasize. In this regard the student Ellie is represented as far more 'developed' than her tutor, Prof Ptthmllnsprts, who declares that water-babies cannot exist on principle. Behind the heavy-handed sermonizing *The Water-Babies*' view of the authority of science is as complex as its view of gender. Ellie is Tom's reward (he 'gets the girl'), but also his role model and guide (not a passive 'angel in the house' awaiting 'her man'). This fairy tale preaches evolution, but warns its readers against those who might normally be considered evolution's guardians or high priests.

It also offers a series of scientific hypotheses. As Kingsley wrote to Maurice, his aim in writing *The Water-Babies* had been to make children and adults alike understand 'that there is a quite miraculous and divine element underlying all physical nature', and 'that nobody knows any thing about anything, in the sense in wh[ich] they *know* God in Christ, and right and wrong'. 'The Physical science in the book is *not* nonsense,' he insisted, 'but accurate and earnest, as far as I dare speak yet.'[12] It is time to look more closely at Kingsley's 'natural theology of the future', as revealed in his historical, theological and natural historical writings. Though Kingsley's readers would, in time, see these genres of writing as representing distinct and sometimes conflicting aspects or phases of his career, in fact they resolve into a surprisingly coherent whole, if we are prepared to follow Kingsley as he makes connections between history and natural history.

Superfecundity was one of the engines that drove Kingsley's natural theology forward. Competition was 'a universal law of living things', he reminded an audience of trainee priests at Sion College in 1871, and 'physical science' was proving that races were not all the same, demonstrating 'how the more favoured races . . . exterminate the less favoured'.[13] According to what Kingsley proudly called 'my degradation theory', races disappeared by a sort of reverse evolution, to the point where they were no longer to be counted as human life. Degradation came to fascinate Kingsley around 1860 and became something of an obsession. Superfecundity and degradation helped him indulge his violent streak without guilt. 'Bloodshed is a bad thing, certainly,' he told his students, 'but after all nature is prodigal of human life – killing her 20,000 and

---

[12]Charles Kingsley to F. D. Maurice, n.d. British Library, London. AddMSS 41297, f. 147.
[13]Charles Kingsley, 'The Natural Theology of the Future,' in Kingsley, *Scientific Lectures and Essays* (London: Macmillan, 1890), pp. 313–36 (324).
[14]Charles Kingsley, *The Roman and the Teuton: A Series of Lectures Delivered before the University of Cambridge* (London: Macmillan 1889), pp. 13–14.

her 50,000 by a single earthquake . . .'[14] If the humans in question had 'degraded' through idleness or exhaustion, then they weren't really human and hence not worth getting overly worried about. Kingsley could accept the destruction of such 'idle' or 'exhausted' races as Providential, and join fellow historians like Carlyle and Froude in sneering at what the latter called 'weak watery talk of "protection of aborigines"'.[15]

Migration lay in the Teutons' very nature, it marked them out from other races. 'The Teutons were and are a strange people,' Kingsley observed, 'so strange that they have conquered – one may almost say they are [conquering] – all nations which are alive upon the globe';[16] Any race which found itself in the path of the Teutons faced a choice: to be assimilated or disappear. Assimilation through conquest also fascinated Kingsley. His Teutons only became fit to govern the world when they learnt 'discipline and civilization' from the Roman armies they fought or in which they served as mercenaries. The boyish Teutonic race needed direction. Happily this race was uniquely gifted in being able to take on a different race's positive characteristics without diluting its own essence in any way. It had that 'teachableness and wide-heartedness, which has enabled us to profit by the wisdom and the civilisation of all ages and of all languages, without prejudice to our own distinctive national character'.[17]

Kingsley's fascination with assimilation by conquest was such that it led the English to change how they viewed their past, making the Norman Conquest of 1066 the key turning point, rather than high political events like the Glorious Revolution of 1688. Macaulay had given the whole eleventh century less than a page in his *History of England* (1848–55), preferring to focus on high political events like 1688. The Conquest had traditionally been represented as a case of a foreign invader hanging his 'Norman yoke' around the necks of free Saxons. Kingsley as well as other assimilationist historians such as Edward Augustus Freeman presented it more as a case of conquest in reverse. In *Hereward the Wake* the faithful Teutonic Saxons lose the battle against the invader, yet also live to fight another day, conquering their conquerors by assimilating with them and instilling Saxon ideals within the Normans. Conquest and death are unmasked as assimilation and rebirth. But can one be an assimilationist and also a racist? Their position seems fatally flawed by internal paradox.

Kingsley, however, found a way out by means of another concept familiar from evolutionary biology: recapitulation. This was the means by which God gave His creatures a second chance, and a third, and a fourth. If they somehow 'failed' the moral test set them by evolution, they went down a class, as it were, and given remedial tuition followed by a resit. As we have

[15]J. A. Froude, 'England's Forgotten Worthies,' in Froude (ed.), *Short Studies on Great Subjects*, 1st series, 2nd edn (1867), pp. 294–333 (305).
[16]Kingsley, *Roman and Teuton*, p. 54.
[17]Kingsley, *Westward Ho!*, pp. 178–9.

seen, Tom in *The Water-Babies* as well as Alton in *Alton Locke* fail, are demoted to lower forms of life, only to climb back up the evolutionary ladder by grit, determination and the encouragement of a good woman, who appears at key points in the story, praising or censuring the creature they observe.

At this point another problem emerges, that of personality. How can we call these various creatures 'Tom' or 'Alton' when their morphology is so utterly different from that of boy and man? It is only possible if we accept that whatever makes 'Tom' 'Tom' and 'Alton' 'Alton' survives all these transformations intact, or, rather, intact *and* improved by all the experience and wisdom 'it' has gained. Kingsley's belief in the transmigration of souls (metempsychosis, i.e. the movement of a single soul from one physical body to another) implied a unitary personality or soul. This metempsychosis had implications for mankind's future state, as well as for its past and present. It confirmed Kingsley's refusal to believe in hell and eternal damnation, in a steady state of suffering ordained by divine judgement. In a way Kingsley's belief in recapitulation and metempsychosis implicated him in a constant deferral of final judgement of an immaterial yet eternal soul. Instead his God prefers regular chastisement, punishment inflicted on a kind of soul that is constantly bleeding from the spiritual into the physical realm. This soul could be found in all forms of life, not just in human beings.

On one level Kingsley's view of hell and heaven was in sympathy with a broader shift in Victorian Christians, one which affected all but a few small sects. Broadly speaking this saw a shift from an Evangelical vision, with hell-fire-and-brimstone on the one side and abstract bliss on the other, to one which found priestly recourse to hell-fire embarrassing, one which saw heaven as a happy setting for relaxation and polite family reunions (a heavenly parlour, in other words). From a doctrinal, church perspective, however, Kingsley was unorthodox, verging on the pantheist. Christian orthodoxy holds that only humans have souls. For them the soul was (and is) the key distinction between humans and the rest of created life. According to Genesis, humanity was created in God's image. The soul is that divine image in humanity. The fact that it remains after our mortal bodies pass away makes a truly long-term relationship with God possible.

By granting everything a soul Kingsley removed the 'big ditch' separating men from apes. He could accommodate and even welcome Huxley's heresies about having an ape for a grandfather. When Huxley wrote Kingsley in May 1863 challenging him on just this point Kingsley was careful to set him straight:

> I never said men had souls and apes had not. I sh[oul]d rather put it – that souls had men, than men souls: but be that as it may, I have every reason to suppose that an ape has a soul, if a man has one; and every other being

> or organized thing – only of a lower organization, according to its degree. There is not a word in Scripture . . . wh[ich] denies *that*.[18]

For Kingsley the body is not a repellent or insignificant husk, and certainly not a sinful tempter, as many Christians believed. Saint Paul had seen a battle going on between a virtuous 'law of mind' and a 'law of sin which dwells in my members [i.e. my limbs]' (Romans 7:23). Kingsley did not share this view.

Instead the body was a faithful representation of the soul within it, and its limbs or 'members' were the tools with which the soul fitted itself for its next life (one of many – there was no longer a single Afterlife-with-a-capital-'A') in a different body. 'Souls secrete bodies,' as Kingsley put it in a letter to the physiologist George Rolleston.[19] This could be seen as 'inheritance of acquired characteristics', as Lamarckianism. This relationship between moral habits and physiological traits can also be found in later writings by Herbert Spencer. 'Character' influenced 'habit', which in turn shaped the nervous system. In an 1876 essay in the journal *Mind*, Spencer could thus speak of 'the cumulative effects of habit on function and structure'.[20]

In an 1879 essay entitled 'The Soul, and the theory of evolution', the science writer Arabella Buckley explained how apparently conflicting concepts could be reconciled if evolution was properly understood

> as a compound of inheritance and the accumulated experiences of each new individual. Reminiscence, ancestral likeness, race characteristics, animal passions, the struggle between the higher good, and the lower nature in which mere propensities have become conscious evil when higher possibilities have been developed – all these are explicable on the theory of evolution.[21]

In *The Water-Babies* Tom's various metamorphoses on the road back to humanity are the reward for having developed moral faculties. Several bodies house his unitary soul as it improves.

Buckley's idea of transmigration emerged from her interest in spiritualism, a movement (addressed in Chapter 10) to which Kingsley was almost entirely indifferent. Animal souls and transmigration were not part of the spiritualist mainstream. Kingsley's ideas came instead from his 1849 reading of Pierre

---

[18]Kingsley to Huxley, [23 May 1863]. Imperial College, London. Huxley Papers, Gen. Letters IX (I-K), f. 235.

[19]Kingsley to Rolleston, 12 October 1862. Kingsley, *Charles Kingsley*, 1: 133–4. See also Kingsley to Huxley, 17 [May?] 1865. Imperial College, London. Huxley Papers, Gen. Letters IX, f. 221.

[20]Herbert Spencer, 'The Comparative Psychology of Man,' in Spencer (ed.), *Essays: Scientific, Political, and Speculative*, 3 vols (London: Williams and Norgate, 1891), 1: 363.

[21]Arabella Buckley, 'The Soul, and the Theory of Evolution,' *University Magazine* 3 (January 1879): 1–10 (10).

Leroux and his circle of French socialist thinkers, to whom he was most likely introduced by fellow Christian Socialist J. M. Ludlow. Like Kingsley, they sought solidarity with the workers and reviled those who insisted on obedience to the 'laws' of political economy, to 'laissez-faire' capitalism. They, too, advocated universal manhood suffrage without seeking (as Charles Fourier did) to purge society of family, property and a Christian morality. Like Kingsley, they refused to believe in eternal damnation and instead held that the self went through multiple lives.

As Leroux put it in his *De l'humanité* (1840), 'To live is to die in one form in order to be reborn in another form.'[22] Jean Reynaud's entry for 'Ciel' ('Heaven') in the *Encyclopédie nouvelle* (1836–40) that he edited with Leroux clearly inspired Kingsley:

> Thus the soul, which passes from one journey to another, leaving its first body for a new body, constantly changing its residence and its exterior, pursues under the rays of the Creator, from transmigration to transmigration and metamorphosis to metamorphosis, the palingenetic course of its eternal destiny . . . Birth is not a beginning, it is merely a change of body.[23]

In the context of contemporary French comparative anatomy Leroux and Reynaud were firmly on the side of the transmutationist Geoffroy St Hilaire (who contributed to the *Encyclopédie nouvelle*), not that of Cuvier who believed, not in the limitless and infinitely variable forms of life, but in the idea that all life was grouped around certain model forms or archetypes.

Rather than being deferred until the Apocalypse, God's judgement of humanity was constant. It was uniformitarian rather than catastrophist, a story of ongoing, routine activity, rather than long periods of inactivity interrupted by cataclysmic change. Metempsychosis made this constant judgement possible. Degenerated life forms were paying the price for their misbehaviour in a previous life, but recapitulation held out the promise of redemption. With multiple lives at his disposal Kingsley could reconcile the stern God of the Old Testament, punishing unto several generations, with the all-merciful Christ of the New Testament. God was 'too good an instructor to lose finally any of his pupils'.[24] Though Kingsley was the first to admit that this uniformitarian system of recapitulation, competition and assimilation was a work in progress, the overall aim was clear. He was 'working out points of Natural Theology, by the strange light of Huxley, Darwin and Lyell'.[25]

---

[22]Lynn Sharp, 'Metempsychosis and Social Reform: The Individual and the Collective in Romantic Socialism,' *French Historical Studies* 27(2) (Spring 2004): 349–79 (368).
[23]Cited in Ibid., 369.
[24]Kingsley, *Charles Kingsley*, 1: 57.
[25]Kingsley to Maurice, n.d. [1863]. British Library, London. AddMSS 41297, f. 147.

Dogmatic atheism versus agnosticism

Confronted with the onward march of scientific discovery, Christians are often represented as facing a choice: either to advance, investing every new discovery with divine agency, or to retreat, appealing to those parts of the universe that geologists, biologists and other men of science have yet to explain. The latter strategy, of attributing to divine agency whatever remains unexplained by natural agencies, is sometimes known today as 'the God of the gaps'. Kingsley certainly saw a clear choice, but on different terms. As he put it to the entomologist Henry Bates in 1863, 'a convert to Darwin's views', *could* view the world as being 'like an immensely long chapter of accidents', but it was 'really . . . a chapter of special Providences of Him without whom not a sparrow falls to the ground'.[26]

Crucial to this view was a distinction between primary and secondary laws governing the universe, between the conventions which described the normal behaviour of living things and the overarching divine Providence which directed the action of secondary laws in certain directions. Though distinct, these two were inseparable. Christians were not faced with a choice between a constantly meddling God and a 'watchmaker' who set things up and then let the universe run by secondary laws, but between a sustaining God (present 'in' everything, including secondary laws) and Godless 'chance'. To follow the maternal metaphors from his 1868 *Good Words* serial (published as *Madam How and Lady Why* a year later), natural science ('Madam How') was concerned with discovering secondary laws, while theology and metaphysics ('Lady Why') were concerned with overarching purpose. Madam How was the 'servant', Lady Why the 'mistress'.

In the 1863 correspondence quoted in the introduction Huxley seemed willing enough to agree with Kingsley. It was the natural historian's duty to break down those doctrines of natural science which failed to withstand experimental testing, but not to construct new supernatural explanations or laws. 'I believe in [William] Hamilton and Herbert Spencer so long as they are destructive and I laugh at their beards as soon as they try to spin their own cobwebs,' Huxley wrote, referring to two of Britain's leading nineteenth-century metaphysicians. 'Is this basis of ignorance broad enough for you?'[27] In the 1870s Huxley coined the word 'agnosticism' to explain this position, which he saw as the most 'scientific' one. Meanwhile Kingsley could write to Maurice claiming that recent publications by David Ansted (geologist), Asa Gray (botanist), Sir William Grove (physicist) and Lyell

[26]Kingsley to [Henry Bates], 13 April 1863. Princeton University Library, Princeton, NJ. AM18153.

[27]Huxley to Kingsley, 22 May 1863. Imperial College, London. Huxley Papers, Gen. Letters IX, f. 229. Reproduced in Leonard Huxley, *Life and Letters of Thomas Henry Huxley*, 2 vols (London: Macmillan, 1900), 1: 242–4.

(geologist) indicated a growing consensus among 'strictly scientific men' in favour of 'a living, immanent, ever working God'.[28]

There was a danger, Kingsley recognized, that the men of science would get carried away and claim to explain everything by means of laws – laws which left no room for divine agency, or individual agency, for that matter. They might fetishize their laws. They might even come to insist that nobody but gentlemen of science were qualified to write or think about them, establishing a kind of scientific priesthood. A church scientific might emerge, one eerily similar to the Roman Catholic church in its tendency to interdict or excommunicate those who refused to accept without question certain doctrines as well as the authority of priestly intermediaries. In 1860, even as he hailed Darwin's discovery, Kingsley expressed such fears in a letter to Huxley. 'I fear, at times, a bigotry of science, and a narrowness of science.'[29]

By 1872 Kingsley felt that Huxley's allies, the physicist John Tyndall and the philosopher Herbert Spencer, were taking their scientistic bigotry too far; the former in an article he wrote on prayer for the *Contemporary Review* and the latter in an article on the journalist and author Harriet Martineau, also in the *Contemporary*. 'They profess not to be searching,' he wrote to an unidentified correspondent, 'but to have found; and are so assured that they have found, that they are becoming aggressive, and will not leave quiet folks in peace.' Not only aggressive, he continued, but also conceited and arrogant:

> There are many men – I among them – who love physical science as dearly as Spencer or Tyndall can; who are ready to follow Darwin, and the "revolution" doctrines, as far as we see good, without the least fear. But when it is said to us. No. You shall not be a scientific man and a Xtian [Christian] you shall not even be one and a Deist. If you make any pretension to science you must join us, repudiate the possibility of metaphysic, repudiate all past philosophies (save Cynicism) and theologies; and believe that no one even knew anything about the human race till French Atheism and Materialism arose 100 years ago; and join us in smiling pityingly at all the noblest . . . interests of humanity: then, I think, an honest man has a right to lose his temper deliberately, and use a few hard words:[30]

Tyndall's address to the 1874 meeting of the BAAS on the conservation of energy, speciation and consciousness certainly led many to wonder if science

---

[28]Kingsley to Maurice, n.d. [1863]. British Library, London. AddMSS 41297, f. 147.
[29]Kingsley to Huxley, 26 September 1860. Imperial College, London. Huxley Papers, Gen. Letters IX, f. 180.
[30]Kingsley to Unknown Correspondent, 9 July 1872. Princeton University Library, Princeton, NJ. AM14756.

was becoming arrogant as well as materialist. 'We claim,' Tyndall thundered, 'and we shall wrest from theology, the entire domain of cosmological theory.' 'All schemes and systems which thus infringe upon the domain of science must, in so far as they do this, submit to its control, and relinquish all thought of controlling it.' As Tyndall's ally Huxley had feared, many failed to appreciate the distinction between crude materialism and Tyndall's 'Higher Materialism', which referred to 'latent powers' described as 'the manifestation of a Power absolutely inscrutable to the intellect of man'.[31]

Part of this 'conceitedness' lay in overestimating the import of scientific 'laws', making them out to be proscriptive, rather than simply descriptive, to claim that they predicted the future, that laws were destiny, that their action could not be interrupted or directed. Ever the sanitary campaigner, Kingsley took a familiar example: cholera spreads, infects and kills in a set, conventional way. It isn't carried by air one day, only to spread by water the next. But we do not see those who die as 'foreordained' to die and refuse to improve water supplies. Or rather, we should not do so: Kingsley's 'Chartist' credentials lay precisely in a hatred of utilitarian political economy (the kind popularized by Martineau's writings). Kingsley claimed that talk of Malthusian or other 'laws' of political economy encouraged moral inertia. Poor law guardians and employers were supposedly led to shrug their shoulders whenever they encountered slums, disease and foul working conditions, to stand aside and do nothing to help, insisting that the 'laws' of the market had to run their course uninterrupted. This is the attitude Dickens attributes to Scrooge in *A Christmas Carol* (1843); the attitude that refuses to help the poor because they represent excess population doomed to die.

In the 1870s and 1880s Huxley seems to have shifted from a 'descriptive' towards a 'prescriptive' (laws as 'imperatives') view of 'law', along with Tyndall, Spencer and others. Kingsley died in 1875, but there were others prepared to take his mantle and try to keep these men from getting carried away. Among these were a Liberal politician and biologist, the Duke of Argyll, the Roman Catholic comparative anatomist, St George Jackson Mivart, and, in the United States, Asa Gray. All welcomed Darwin, yet co-opted transmutation into a model of 'evolution by law' in which God's sustaining presence was constant. All made useful criticisms of the scientism of Huxley and his allies and hence were airbrushed out of twentieth-century histories of science, despite the fact that by the 1890s their views represented the consensus.

The key texts for 'evolution by law' were Argyll's *Reign of Law* (1867) and Mivart's *On the Genesis of Species* (1871). Both pointed to opposable

[31]Tyndall, 'The Belfast Address,' in Tyndall (ed.), *Fragments of Science*, 6th edn, 2 vols (London: Longmans, Green, 1879), 2: 191–3. See also Lightman, 'Scientists as Materialists in the Periodical Press: Tyndall's Belfast Address,' in Lightman (ed.), *Evolutionary Naturalism in Victorian Britain: The 'Darwinians' and their Critics* (Farnham: Ashgate, 2009), ch. 5.

thumbs, eyes, brains and the faculty of speech as features which natural selection could not explain, which involved the parallel emergence of distinct features, none of which in isolation were of any benefit to the creature. Some Higher Intelligence must, therefore, have been guiding transmutation in a particular direction. To Argyll laws became instruments through which God manifested His power over the universe, 'essential implements or tools in the hands of Will'.[32] Mivart took this 'Will' even further, proposing 'that all force may be will force; and thus, that the whole universe is not merely dependent on, but actually *is*, the WILL, of higher intelligences, or of one Supreme Intelligence'.[33] The spiritualism of Alfred Russel Wallace (discussed in Chapter 10) welcomed such language. Unlike natural selection, guided 'evolution by law' coped well with the recalculation of the earth's age unleashed by the work of the physicist Lord Kelvin (which suggested that the earth was not in fact millions of years old). Admittedly, discerning exactly how these different types of 'laws of creation' are supposed to work can be difficult. As Peter Bowler has noted, they were popular precisely because they fudged issues.[34]

In the narrower context of palaeontology Argyll went on to propose a principle of degradation in works such as *Primeval Man* (1869), one that balanced regression against progression. 'Man's capacities of degradation stand in close relation,' Argyll wrote, 'and are proportionate, to his capacities of improvement.'[35] Whether as a 'law' or a 'principle', 'degradation' or 'degeneration', Argyll's concept grew in stature in the years after Kingsley's death. The German zoologist Anton Dohrn advanced his own 'degeneration principle' to Darwin, arguing that it had to be incorporated in Darwinian zoology. Alfred Russel Wallace came out in support at the 1876 meeting of BAAS, and 4 years later the zoologist Ray Lankester published his *Degeneration: a Chapter in Darwinism*, which showed how creatures presented with a glut of food degenerated, just like the lazy Doasyoulikes do in *The Water-Babies*. From the 1880s onwards degradation became less of a brake, intermittently interrupting evolution's upward progress, and more like a gravitational force, a force that progressive evolution struggled in vain to escape.

For Christian evolutionists like Kingsley, Darwin and his 'strictly scientific men' had brought us closer to understanding God's power as something dynamic that invested all of creation all of the time. In replacing a catastrophist, saltationist model with a uniformitarian, steadily evolving one, men of science had unwittingly dispensed with God as 'master-magician'. Both this 'master magician' and his worshippers were demeaned, Kingsley

[32]Agyll, *What is Science?* (Edinburgh: David Douglas, 1898), p. 29.

[33]George St Jackson Mivart, *On the Genesis of Species* (London: n.p., 1871), p. 280.

[34]Peter J. Bowler, *Evolution: The History of an Idea,* revised edn (Los Angeles: University of California Press, 1989), p. 144; Peter J. Bowler, *Non-Darwinian Revolution: Reinterpreting a Historical Myth* (Baltimore: Johns Hopkins University Press, 1992), p. 63.

[35]Argyll, *Primeval Man: An Examination of Some Recent Speculations* (London: Strahan, 1869), p. 192.

argued, every time the latter called on the former (e.g. during official fast days) to act 'on' or 'interrupt' nature as if interacting with something outside Himself.[36]

In moving away from Christian Socialism Kingsley also drifted apart from his 'master', F. D. Maurice. But Kingsley himself saw it differently, arguing that the Church of England had a proud tradition of combining philosophy and natural history with theology. In 1846 Maurice refused an invitation to make *Vestiges* the focus of his Boyle Lectures. To Kingsley such an action, though born of Maurice's awareness of his own ignorance of the natural sciences, was a derogation of the churchman's responsibility to engage with and comment on the latest scientific developments. Evolution had made it possible to reconstruct natural theology in a more awe-inspiring and therefore truer fashion. In 1860 Kingsley wrote to Huxley that

> . . . there comes over me at times a vision, wh[ich] I would not sell for all the gold on earth, of infinite and perpetual upward development, going on all around from all ages to all ages, of the inorganic into the organic, of the organic into the animal, the animal into rational, the rational into that higher sphere of being for wh[ich] as yet we have no name. . . . I may have developed from a monad: but I will not be bound down by the conditions of the monad, in my thoughts and aspirations about myself, because I know as a fact that I have quite infinitely greater capacities than it. I dare say that I am descended from some animal from whom also the chimpanzee has sprung – I accept the fact fully, and care nothing about it . . . . if the laws of nature (customs of matter, I call them) can kill the chimpanzee, they cannot – I had almost said they shall not – kill me.

Kingsley and his fellow men would live again, 'the greater by all that we have learnt in this life', and 'begin again where we left off at death – I trust to learn new lessons, in a new schoolhouse'.[37]

Kingsley revelled in watching Darwin's ideas spread, and Darwin himself recognized that he had a useful ally in Kingsley. But Kingsley clearly did not follow Darwin religiously. After a merry house party at Lord Ashburton's spent discussing Huxley and Darwin's ideas with the Duke of Argyll and the Bishop of Oxford, Samuel Wilberforce, Kingsley wrote to Huxley reiterating his delight at Darwin's fame, but added an important caveat: 'You cannot conceive how Darwin's views are spreading – with – of course, demurrers and reconsiderations, quite necessary in so great and new a vista of thought.'[38]

---

[36]Frank Turner, 'Rainfall, Plagues and the Prince of Wales,' in Turner (ed.), *Contesting Cultural Authority: Essays in Victorian Intellectual Life* (Cambridge: Cambridge University Press, 1993), p. 155.

[37]Kingsley to Huxley, 21 September 1860. Imperial College, London. Huxley Papers, Gen. Letters IX, f. 162.

[38]Kingsley to Huxley, 28 February 1862. Imperial College, London. Huxley Papers, Gen. Letters IX, f. 203.

Darwin had served science and faith by stirring up speculation, by asking more questions than he solved. For Kingsley good science was that which replaced comforting certainties – which usually turned out to be wrong – with unsettling hypotheses that multiplied mysteries.

When Lord Kelvin sent him a copy of his work recalculating the age of the earth Kingsley replied, noting his gratitude to Kelvin for keeping his mind 'in a state of healthily chronic change'.[39] Neither the life of science nor the life of faith was meant to be easy. But 'chronic change' was 'healthy', a sign of life. Truth was not a static target, but, like the horizon, always retreating. Science demanded constant activity, a constant running to keep up. This focus on manly activity and courage in the face of obstacles, celebrated almost for their own sakes, was central to Kingsley's character. It also makes him quintessentially Victorian.

Further reading

Beer, Gillian, *Darwin's Plots: Evolutionary Narrative in Darwin, George Eliot and Nineteenth-Century Fiction* (Cambridge: Cambridge University Press, 2000).

Conlin, Jonathan, 'An Illiberal Descent: Natural and National History in the Work of Charles Kingsley,' *History* 96 (April 2011): 167–87.

Helmstadter, Richard J. and Bernard Lightman (eds), *Victorian Faith in Crisis* (Basingstoke: Macmillan, 1990).

Kingsley, Fanny, *Charles Kingsley: His Letters and Memories of His Life,* 9th edn, 2 vols (London: Macmillan, 1881).

Klaver, J. M. I., *The Apostle of the Flesh: A Critical Life of Charles Kingsley* (Leiden: Brill, 2006).

Maynard, John, *Victorian Discourses on Sexuality and Religion* (Cambridge: Cambridge University Press, 1993).

Merrill, Lynn L., *The Romance of Victorian Natural History* (Oxford: Oxford University Press, 1989).

Moore, James R., *The Post-Darwinian Controversies. A Study of the Protestant Struggle to Come to Terms With Darwin in Great Britain and North America, 1870–1900* (Cambridge: Cambridge University Press, 1979).

Norman, Edward, *The Victorian Christian Socialists* (Cambridge: Cambridge University Press, 1987).

Turner, Frank, *Contesting Cultural Authority: Essays in Victorian Intellectual Life* (Cambridge: Cambridge University Press, 1993).

Uffelman, Larry K., *Charles Kingsley* (Boston: Twayne, 1979).

Vance, Norman, *The Sinews of the Spirit: The Ideal of Christian Manliness in Victorian Literature and Religious Thought* (Cambridge: Cambridge University Press, 1985).

[39]Kingsley to William Thomson, 15 June 1869. Princeton University Library, Princeton, NJ. AM85–73.

CHAPTER SIX

Imperial evolution? 'Greater Britons' and other races

On 8 February 1886 the members of Macclesfield's Townley Street Mutual Improvement Society sat down to listen to a lecture on 'The Making of Greater Britain', delivered by Henry Birchenough (1853–1937). Birchenough set himself to explain to his audience how they had come to control the largest empire the world had ever seen, and to remind them of the responsibilities that came with it. 'How was it that we passed from the condition of an island people to that of a world-pervading race, dwelling as naturally in Canada, in South Africa, or in Australia, as in Cheshire or Kent?'[1] Compared with Portugal and Spain, Birchenough noted, Britain had left her imperial expansion rather late. Many of his audience, he assumed, had read Kingsley's *Westward Ho!* and so knew all the stories about Admiral Drake and the Spanish Armada. From these beginnings, he noted, the English had built a new kind of empire, one based on largely peaceful settlement and free trade, rather than bloody conquest.

This empire was a community of race, one that transcended political borders such as those that marked out the former American colonies as an independent United States of America. The one exception – a very large one – was India. Britain's Indian empire 'has always represented conquest, and not true expansion; . . . . India is not properly a part of Greater Britain at all. It lies outside it as a colossal appendage to the English-speaking empire'.[2] The British Empire was also 'on the side of liberty and progress, of justice and humanity'. Here the exception was Britain's massive involvement in the

---

[1]Henry Birchenough, *The Making of Greater Britain* (Macclesfield: Claye, Brown and Claye, 1886), p. 4.
[2]Ibid., p. 23.

slave trade, Birchenough noted. Even after the trade in slaves was abolished in 1807, slave keeping remained legal in Britain's Caribbean colonies until 1833. 'Her hands were more deeply stained with slavery than those of any other people,' Birchenough noted.[3]

'Greater Britain' represented more than a glorious past, however. This empire had continued relevance to his Macclesfield audience, to 'England' as she faced today's challenges: the rise of Prussia, overcrowded slums, working-class unemployment, localized violence on her imperial frontiers, such as the Mahdi rising in the Sudan, which had recently claimed the life of Major-General Charles Gordon, the governor-general. 'Greater Britain' provided an outlet for overcrowding at home, a place to emigrate to and find employment. 'Greater Britain' had nothing to fear from either the rise of industrialized Prussia, Birchenough insisted, or from the rising powers of Russia and the United States. This was because of 'Greater Britain''s unique advantage: 'practically all her lands are inhabited by people of the same race.'[4] This had been demonstrated in the wake of Gordon's death by the many condolences and offers of military support which had been relayed 'home' by telegraph from English settlers in New Zealand, Canada and countless other English-speaking territories straddling the globe. Along with a glorious past and present, therefore, 'the history of the future', too, was one in which Britain took the starring role.

The speaker would have been well known to his audience. His family had deep roots in this northern industrial city, where they ran a silk works, and shared its Liberal, non-conformist civic culture. The family were Methodists, Henry's father had served as mayor and Henry himself served as president of the city's Chamber of Commerce. At Oxford Henry had become friends with the future imperialist Alfred Milner, who got him involved in the New Liberal project of creating an 'Imperial Federation', that is an imperial free-trade area. Henry served as director of the Imperial Continental Gas Corporation and was later sent to South Africa by Milner as Trade Commissioner to assess business opportunities for Britain in the wake of the Boer War. He also chaired select committees on matters relating to industrial competitiveness and cotton growing in the empire.

Though the lecture was published locally as a pamphlet, it might seem an odd way to begin a chapter on the British Empire's role as an evolutionary laboratory. Neither the setting, the occasion nor the speaker are particularly memorable. Granted, Birchenough's career clearly shows how the fabric of 'Greater Britain' was quite literally woven from threads linking business and government, peace and war, capital and provinces, agriculture and industry. But he was hardly exceptional in that. He does not rate an entry in the *Oxford Dictionary of National Biography*. His lecture is worth quoting not because of any originality but because of its representativeness. It represents

---

[3]Ibid., pp. 7, 25.
[4]Ibid., p. 27.

the consensus view of empire held by most educated Britons between about 1870 and 1914. As we are already beginning to appreciate, this view was founded on notions of race and what Birchenough called 'uninterrupted progress and development'. Evolutionary notions.

What is 'race'? The word had been used to refer to an individual's family or kindred since the sixteenth century, although this meaning was becoming archaic in the Victorian period. It had also been used to refer to a tribe or nation. The idea of 'race' as centred around a group's shared language, customs or culture developed in the late eighteenth century. In the course of our period 'race' shifted from being a descriptive term to a classificatory one, from a focus on languages and culture to morphology. Ranking by levels of 'civilisation' and sometimes intelligence formed part of this classification. This shift can be seen in the shift in the labels used to refer to the study of race: 'ethnology' and 'ethnography' gave way to 'anthropology.' Victorian practitioners are often described as 'monogenist' or 'polygenist', depending on whether they believed in the 'unity of man' (that all races were one 'common stock') or saw races as distinct.

From an evolutionary perspective this is a distinction without a difference. All humans, indeed, all life is derived from one common stock. It was nonetheless possible for Victorians to accept evolution and nonetheless hold that different 'races' (understood as varieties of *Homo sapiens*) had divided such a long time ago that the descendents of each line were as good as separate species. Monogenists could concede a single line of descent for different races, yet describe this or that race as 'arrested' (i.e. as one that had stopped evolving at a certain point, being left behind) or 'degraded' (i.e. one that it had fallen behind its 'peers'). The discovery of Neanderthal and other fossil hominids from the 1850s onwards served to provide a set of markers to which contemporary 'savages' could be referred to by analogy, further confirming the implied shame in having stopped or gone into reverse.

For many Enlightened thinkers the discoveries and contacts resulting from the empire provided a new perspective from which to question their own moral and cultural values. Writing in 1759, Adam Smith praised the stoicism and magnanimity of the Iroquois and West African slave, declaring both to be beyond the comprehension of the Europeans with whom they came into contact.[5] By the early nineteenth century, however, an unspoken, unexamined belief in the superiority of the colonizer over the colonized had taken hold. Depending on how one understood evolution, the responsibilities of empire could look different, however. The empire could be natural selection writ large, a means by which a leading race exterminated or assimilated all others. It could be an information-gathering network, a kind of racial observatory or even a zoo, protecting races against destruction. It could

[5]Adam Smith, *The Theory of Moral Sentiments*, eds. D. D. Raphael and A. L. Macfie (Oxford: Clarendon, 1976), p. 206.

challenge our unexamined views of 'fitness'. Finally, it could be a source of dangerous moral or physical contagion, threatening the 'fittest' race with extinction, rather than with survival.

Absence of mind?

The British Empire originated not as a royal or state project, but as a private-public partnership. Over the course of the sixteenth and seventeenth centuries groups of merchants, ship-owners and financiers clubbed together to form a series of joint-stock companies, each of which bribed the court to secure a royal charter granting them a monopoly of British trade with a particular part of the world, such as Africa or the Levant (i.e. the eastern Mediterranean), where they wished to set up trading posts or 'factories'. The policy of such companies with regard to indigenous power structures and peoples was clear: no war, no conquest. To quote Thomas Arne's famous patriotic ditty, 'Rule Britannia' (from a masque written in 1740), Britannia was instructed to 'rule the waves'. Ruling the land was what other empires did.

Of course, to present this as a deliberate policy is to suggest that Britain had a choice, when her weakness relative to the Portuguese, Dutch, Spanish and French dictated her choice until the nineteenth century. It is also misleading in that it suggests that someone somewhere was thinking about something called 'British imperial policy' before the middle of the nineteenth century, and very few people were. Adam Smith was one of them, and argued in his *Wealth of Nations* (1776) that Britain should give up her empire. Indeed, he argued that she should do so immediately, without even waiting to see how the American colonists' revolution turned out. Britain fought hard to keep what became the United States, but the shock of defeat in 1781 was got over quickly. In 1784 a political crisis over plans to increase state influence over the East India Company (the EIC, the company that administered India) eclipsed it, as the nation came together in *opposition* to the corruption they associated with imperial administration.

The transition from an 'empire of the seas' to 'Greater Britain' was driven by three factors, all of which nudged a reluctant, sceptical British state to shoulder the high fiscal-military costs of administrating far-flung territories over the closing decades of the eighteenth century and the first half of the nineteenth century. The first was settler-driven and experienced most commonly in North America, Australia and New Zealand. As settlers moved inland in search of land for farming or grazing they and their concept of property came into conflict with indigenous peoples with very different ideas of who could and did control land usage. Restrictions to settlement agreed by treaties negotiated between native rulers and representatives of the British state were ignored by settlers looking for more land and others (miners, trappers, loggers) looking to exploit natural resources. When indigenous people resisted the latter clamoured for the British state to intervene. In some cases settlers described inconclusive frontier skirmishes as 'massacres' committed by 'savages', arguing that the latter should be exterminated or driven into reservations. The most notorious example of this was the reservation established by a missionary named George Robinson in 1831 on Flinders Island, to which native Tasmanians were consigned until the last inmate died in 1876.

The second factor, particularly evident in West Africa and across Asia, was a shift in the balance of power between the trading companies and the states or communities which hosted their factories. Initially this relationship was one of client and patron. In the century after the Royal African Company's capture of Cape Coast Castle from a rival Danish company in 1664 it was clear that the English (or rather, the English company, as opposed to the Dutch, French, Prussian or Swedish ones) remained there only because the indigenous Fanti ruler allowed them to. The military assets available even to the governor of an important factory like Cape Coast were laughable. In the 1760s it was rare for the governor to receive the cannon salute he was officially due whenever he made his ceremonial entry. Not only was there not enough gunpowder, the castle's ramparts had a nasty habit of collapsing on the rare occasions the cannon *was* fired.

The balance of power shifted, however, as indigenous rulers became dependent on the new income streams opened up by the European companies. Whereas the Fanti had been enslaving those they captured in wars with neighbouring tribes for centuries, British demand led them to move their slave taking up several gears. It also changed the nature of slavery itself, shifting it from forced labour to 'chattel slavery', in which the slave was viewed as a form of property (e.g. one that could be left in a will) and transported thousands of miles across the Atlantic (thousands died en route). Loans, military advice and exchange of technology and knowledge further tightened connections between indigenous elites and the English traders.

This fluid and largely symbiotic relationship was complicated by the third factor driving Britain's shift from an empire of trade to 'Greater Britain', that is, competition between European powers for control of trade routes and influence over local elites. The Seven Years War (1756–63) marked the turning point, both for Britain's emergence as a European power and in the history of war itself. With campaigns in Canada, the Caribbean, the Philippines and Africa as well as in continental Europe, it could be called the First World War. Britain and France forged alliances with a variety of Native American tribes and Indian principalities. In many cases indigenous rulers successfully played Britain off against France (or vice versa) or made use of their European allies' forces and expertise to settle their own scores with local rivals.

Insofar as the British state saw treaties and diplomacy as its responsibility, however, this certainly served to increase British state involvement in her trading companies' doings, as a matter of long-term strategy rather than (as before) short-term expediency. This was still a somewhat defensive strategy, however. Whig reformers and free trade campaigners of the 1830s and 1840s continued to oppose imperialism and found the EIC an embarrassment. The 1833 Government of India Act saw the EIC stripped of its trading monopoly. In 1858 another act marked the completion of the British government's slow takeover of the EIC and the territories it controlled. The immediate impetus was a disastrous mutiny (1856–57) by EIC sepoys (i.e. native troops) widely blamed on EIC mismanagement. It was just one of the many Victorian conflicts fought to hold India and the lines of communication linking India and Britain.

'We seem, as it were, to have conquered and peopled half the world in a fit of absence of mind.' At least, according to J. R. Seeley, Kingsley's successor as Regius Professor of History at Cambridge, who described her collecting territories and expanding her sphere of influence in an unconscious, incoherent and uncoordinated fashion.[6] Though contested, this traditional view does acknowledge several key differences between the British and other European Empires. The great Portuguese and Spanish Empires were the products of close coordination between powerful monarchies and the Roman Catholic Church in the fifteenth and sixteenth centuries, and were founded on military conquest. After the Reformation and the Civil War neither the Church of England nor the English crown was in a position to dream up anything similar. Apart from a short-lived eighteenth-century predecessor, there was no Colonial Office to administer the empire until 1854. To describe the British Empire as the result of coordination or planning is difficult. Evolution was welcomed, therefore, partly because it explained how a messy process characterized at different times by aggression, ignorance, greed, philanthropy, cruelty, incompetence and indifference could nonetheless reveal a higher teleology, a British-focused imperial providence.

Enlightenment and Emancipation

In his *Sketches of the History of Man* (1774) the Scottish philosopher Henry Home, Lord Kames had concluded 'beyond any rational doubt, that there are different races or kinds of men'. A beneficent God had given each species of human a 'complexion, features, shape, and other external circumstances' as well as a 'temper and disposition' suited to its climate. Wherever they were placed, men had begun in the savage state and then advanced in 'civilization'. Some had civilized more than others, but this was a question of 'condition', of the physical as well as the social environment in which they found themselves placed. Differences in 'condition' did not imply a difference

[6]J. R. Seeley, *The Expansion of England* (Cambridge: Cambridge University Press, 2010), p. 8.

in 'understanding' or any other human faculty, only the lack of opportunity to develop that 'power'.[7] Kames was not presenting human diversity of origin as tantamount to a diversity of innate abilities (intelligence, say). His 'Negro' and 'White' are equally intelligent, but that does not mean that they are the same species.

The first sustained attempt to explore 'the origin and mutual relation of human races' was undertaken in the early decades of the following century by a Bristol doctor, James Cowles Prichard (1786–1848). After studying anatomy at St Thomas' Hospital in London Prichard moved to Edinburgh University where he completed a doctorate in medicine in 1808. At Edinburgh he became interested in Scottish Enlightenment theories of human development. His doctoral thesis was published as *Researches into the Physical History of Mankind* (1813), which grew in subsequent, heavily rewritten editions to two (1826) and eventually five volumes (1836–47). Prichard's work cited the usual travellers' accounts (as Kames had), but also engaged with continental comparative anatomists who seemed much more discriminating in their scholarship, derived as it was from a close reading of human skulls collected from around the globe. This science of craniometry had been established by the Dutchman Petrus Camper (1722–89) and the German Johann Friedrich Blumenbach (1752–1840). Blumenbach's craniological researches had assigned a distinct shape of skull to five distinct races: Americans, Ethiopians, Caucasians, Mongols and Malays. Camper identified a distinct 'facial angle' for Europeans, Africans and other races. The more the forehead sloped backwards, the lower the angle.

By assigning a perfect 90 degrees to an ancient Greek statue (then considered the highest form of art), 80 degrees to Europeans, 70 degrees to Africans and 50 degrees to an orangutang, Camper had provided a ready-made ladder with 'favoured' races at the top and less 'favoured' ones at the bottom. Though Blumenbach did not attribute greater intelligence to this or that race, his belief in degenerationism did lead him to conclude that the other races were degenerated forms of a Caucasian Adam and Eve. The early nineteenth century saw a number of scholars from the next generation use such categories to develop a racist ethnology. Among them was the French natural historian Jean Baptiste Bory de Saint Vincent, in his *Zoological Essay on the Human Species* (1827).

Prichard described the aim of his *Researches* as being to establish 'whether two races of animals or of plants, belonging to the same genus and similar in many respects but different in others, are in reality so many distinct species or merely varieties of one species'. Prichard identified two lines of enquiry, one of which he considered 'analogical', based on investigations of how 'the laws of the animal economy' caused 'physical characters and

---

[7]Henry Home, Lord Kames, *Sketches of the History of Man*, ed. James A Harris, 3 vols (Indianapolis, IN: Liberty Fund, 2007), 1: 36.

constitution' to vary. This study explored what variations could occur, and so he compared the fertility, longevity, psychology and development (as in, from childhood to maturation) of 'Africans' with those of 'Americans'. He found few differences; African women did not breed faster or more frequently than European ones. All races exhibited 'a reference . . . to a state of existence after death', expressed in buildings, priestly castes, pilgrimages and so on. One wonders what his readers made of having their Christian beliefs and sacred rites described as 'psychical phenomena' (Prichard asked permission to use this phrase).[8]

When it came to skulls, Prichard was willing to concede that their morphology was the most promising area for those endeavouring to identify distinguishing 'marks' separating races. But he noted, correctly, that the relationship between facial angle or brain size and intelligence had not been made. As for Blumenbach's classification of skulls by breadth and length, this was too flimsy a basis on which to work. The range of morphological variability was so wide that distinct physical 'characters' were far more difficult to find than their work suggested. Prichard proposed three skull types, but did not assign one to this or that race, or link any of them to a given level of intelligence. He concluded that such differences of morphology were a product of variability. Though they might be seen as characteristic of a particular 'tribe', these differences were merely 'varieties', rather than 'specific differences' that could be assigned to one of a number of human species.

Later volumes turned to the second approach, which was 'historic', focusing on what physical changes actually had occurred. Here Prichard's 'ethnography' considered political and military history, religious beliefs and, in particular, language. Whether considering Persian creation narratives, the history of Japan or the claims of a 'Syro-Arabian people' to be the first to inhabit Cappadocia, Prichard's scholarship was wide-ranging, exhaustive and respectful in its treatment of far-flung peoples and their heritage. It was primarily descriptive rather than analytical, presenting facts, but rarely attempting to systematize or classify.

In 1807 parliament had prohibited the trade in slaves, responding in part to the abolitionist movement led by William Wilberforce (1758–1833). This campaign had advanced a number of arguments against human trafficking and slavery. Slave trading and holding were represented as the cause of moral and physical degradation in both slave and master. The informal rules under which black slaves and mixed-race children were maintained by plantation owners and Royal Africa Company 'factors' were taken as evidence that sexual depravity had become ingrained in such communities. Abolitionists also argued that slaves were less economically efficient than free labour, and

[8] J. C. Prichard, *Researches into the Physical History of Man*, 4th edn, 5 vols (London: Houlston and Stoneman, 1851), 1: 176.

claimed that slave trading and the sugar plantations were too dependent on state support to make them viable.

The slogan 'Am I not a Man and a Brother?', which appeared on abolitionist literature (and Wedgwood pottery, thanks to Josiah Wedgwood's support), partly reflected a new humanitarianism, that is, a belief that the welfare of our fellow humans is worth preserving for its own sake, itself related to an idea of mankind as having a special 'dignity'. The word 'humanitarian' did not exist until 1850 (*OED* informs us), and even then was often used pejoratively, to refer to sentimental gush. In the previous century Adam Smith, his tutor Francis Hutcheson and others had debated whether such 'universal benevolence' was too unfocused to be a motivating passion for any individual. One could not sympathize with large groups of strangers living a long way away. It was just as well, therefore, that the abolitionist slogan had divine authority, echoing Jesus Christ's instruction to his disciples to treat one another as brothers (Mark 3:33–4).

Prichard grew up and practised medicine in a port (Bristol) that depended on the slave trade for its prosperity. His father's iron may well have been carried to West Africa (perhaps in the form of shackles) and exchanged for slaves, one of a number of goods (glass beads, guns, woollens, cotton cloth and metal ingots) specifically produced for the African market. Yet he was an abolitionist, if a quiet one. In discussing the languages, skin colours, and religions of West Africa in the second volume of the *Researches* Prichard described the slave trade as a 'diabolical traffic which our legislature, after maintaining for centuries its lawfulness, has, through the growing influence of Christianity on public opinion, at length proscribed'.[9]

Elsewhere he welcomed the success of missionaries in 'civilizing' the Bushmen of South Africa, arguing that such stories represented further evidence that even 'the rudest savages' had propensities to veneration and to moral feelings already within them, without which any attempt to improve their 'outward condition and prosperity' would have been in vain.[10] He presumably shared the hopes of Baptist missionaries active in Jamaica in the 1830s, during that transitional phase of 'apprenticeship' which had been enforced by the 1833 Parliamentary Act abolishing slavery. These had a vision of freed slaves becoming self-supporting tenant farmers, tending plots rented to them by their former owners, wearing 'decent' (Western) clothing, raising a limited number of children within Christian marriage and attending church twice on Sundays.

Similar hopes motivated the missionary George Robinson in his administration of Flinders Island, an enclave in which just under 200 Tasmanians were collected from the 1830s onwards, ostensibly for their own protection. 'I imagined myself an Aborigine,' Robinson wrote, 'I looked

[9] Ibid., 2: 91.
[10] Ibid., 1: 183.

upon them as brethren not, as they have been maligned, savages . . . God has made of one blood all nations of people.'[11] The Aboriginal Protection Society (APS) established by the Edinburgh-trained physician and Quaker Thomas Hodgkin and the Quaker MP Thomas Fowell Buxton in 1837 had as its motto '*ab uno sanguine*' ('from one blood'). At a time when many of Hodgkin's fellow nonconformists, including Spencer, saw the Empire as simply an exercise in violent oppression intended to provide flashy uniforms for aristocratic loungers, the APS saw its potential educative function. In both cases tensions grew between abolitionists, missionaries and other self-appointed 'protectors' of indigenous peoples on the one hand, and planters, ranchers and settlers on the other, who reviled the former as unmanly troublemakers.

THE MORANT BAY REBELLION

On 11 October 1865 a large crowd of several hundred blacks marched into the Jamaican town of Morant Bay and raided the police station for weapons. Tensions in the island between the black population and the colonial government had been high for some time, fuelled by accusations from a member of the Baptist Missionary Society that the government had failed to protect the black population from the effects of a downturn in the island's economy, which had struggled ever since the removal of trade duties protecting sugar. The crowd found the white civilian militia drawn up to protect a meeting of the parish vestry (local government committee). A confrontation ensued, which left 25 dead. In the days that followed bands of black residents roamed the area intimidating plantation owners. Believing these events to be part of an island-wide conspiracy to wipe out the white minority population, the governor of Jamaica, Edward John Eyre (1815–1901), called out the troops and declared martial law in the eastern half of the island.

Eyre's forces struck hard, burning down hundreds of houses, dispensing floggings and causing the deaths of around 500 black Jamaicans. The summary trial and execution of a black former magistrate and Baptist, George William Gordon, was particularly worrying; Gordon was tried and executed under martial law at Morant Bay, even though he had been arrested in the west of the island where martial law was not in force. Eyre, it seemed, was settling an old score with Gordon, with whom he had clashed shortly after Eyre's arrival in 1862. A Royal Commission established to investigate the 1865 rebellion found that Eyre had acted hastily. He returned to Britain in 1866, his career in the colonial service finished.

Opinion in Britain was divided between those led by J. S. Mill, who wished to prosecute Eyre for murder, and the Eyre Defence Committee, which argued

[11]Cited in Patrick Brantlinger, *Taming Cannibals. Race and the Victorians* (Ithaca: Cornell University Press, 2011), p. 55.

that he was a hero who had protected white women and children from rape and murder at the hands of a horde of blacks, using the only language they could understand (i.e. violence). Carlyle and Kingsley, unsurprisingly, were behind Eyre, as was John Tyndall. Huxley, Tyndall's best friend, was on the other side, with Spencer. Huxley wrote to Kingsley that the dispute had revealed an intractable divide in Victorian attitudes:

> In point of fact men take sides in this question not as much by looking at the mere facts of the case – but rather as their deepest political convictions lead them – And the great use of the prosecution and one of my reasons for joining it is that it will help a great many people to find out what their profoundest political beliefs are. The Hero worshipper, who believes that the world is to be governed by its great men – who are to lead the little ones justly if they can; but if not, unjustly drive and kick them the right way – will sympathize with Mr Eyre. The other sort (to which I belong) who look upon Hero worship as no better than any other idolatry: and upon the attitude of mind of the Hero worshipper as essentially immoral: who look upon the observances of inflexible justice as between man and man as of far great importance than even the preservation of social order – will believe that Mr Eyre has committed one of the greatest crimes of which a person in authority can be guilty . . .[12]

Huxley's position of *Fiat justitia, ruat caelum* ('Let there be justice, even if the sky falls') presumed that all humans were equal before the law, including blacks. But this 'doctrine of equal natural rights' could not create equality in intellect where (Huxley insisted) it did not exist. 'No rational man, cognisant of the facts, believes that the average negro is the equal, still less the superior, of the average white man,' Huxley wrote that same year, in his essay 'Emancipation – Black and White'.[13]

One of the ironies of Morant Bay was the fact that Eyre was anything but the stereotypical aristocrat in a crisp uniform. The son of a poor curate, Eyre had emigrated to Australia at the age of 17 and kept a sheep farm. Indeed, his problems on Jamaica and in earlier stints as governor in other British colonies were partly the result of his lowly origins and the social stigma of having been a farmer in Australia (whose origins as a penal colony continue to colour popular perceptions of it as a nation). The white elite resented having such a man lead the status-conscious society typical of imperial outposts, especially when his manners were ungainly.

---

[12]Huxley to Kingsley, 8 November 1866. Imperial College, London. Huxley Papers, Gen. Letters IX, f. 243. Reproduced in Leonard Huxley, *Life and Letters of Thomas Henry Huxley*, 2 vols (London: Macmillan, 1900), 1: 281–2.

[13]The Huxley File. http://aleph0.clarku.edu/huxley/CE3/B&W.html (accessed 13 May 2013).

Having done well as a farmer Eyre had turned explorer, funding his own risky journeys across vast expanses of southern Australia, much of it inhospitable desert. These exploits were 'heroic' enough to inspire Kingsley's brother Henry to write an adventure story about him. Unlike most of his fellow ranchers, Eyre did not believe that the Aborigines were doomed to clash with European settlers and be annihilated. He spent long periods of time with Aborigines and added an ethnological account to the published account (1845) of his explorations. 'The character of the Australian native has been so constantly misrepresented and traduced,' Eyre noted, 'that by the world at large he is looked upon as the lowest and most degraded of the human species. . . . It is said, indeed, that the Australian is an irreclaimable, unteachable being . . . cruel, blood-thirsty, revengeful, and treacherous . . .' In fact, Eyre insisted, the 'worst traits' were the product of custom, while the violent deeds attributed to them were perfectly excusable in the light of the changes that European settlement had introduced to their circumstances. 'Were Europeans placed under the same circumstances, equally wronged, and equally shut out from redress,' he wrote, 'they would not exhibit half the moderation or forbearance that these poor untutored children of impulse have invariably shewn.'[14]

The Aborigines in short were perfectly capable of higher moral feeling and 'civilisation', but only if Europeans took care to understand their ways and work with them, rather than show ignorant opposition to them. Eyre's first posting in the colonial service had been as magistrate and Protector of Aborigines in Murray River (South Australia), a region which had seen regular clashes between Aborigines and settlers. Such clashes ceased entirely under Eyre, who clearly practised what he preached. But Eyre also noticed that wherever Europeans settled, indigenous populations declined in direct proportion. Again, contrary to the consensus among settlers, Eyre insisted that this should not be assumed to be 'the natural course of events . . . ordained by Providence, unavoidable, and not to be impeded'.[15] On the contrary the Europeans had simply taken what was not theirs, illegally, arguing that the lands they enclosed were not being used by aborigines when in fact they supported the animals on which the latter defended for food.

Ethnology or anthropology?

What could have changed Eyre in 20 years from a Prichardian disciple of the 'unity of man' to a man capable of leading the violent suppression of the invisible conspiracy at Morant Bay? The blame for changing attitudes towards race is often laid on the racist ethnology of Robert Knox's book *The*

[14]E. J. Eyre, *Journals of Expeditions of Discovery into Central Australia*, 2 vols (London: T. and W. Boone, 1845), 1: 153, 155–6.
[15]Ibid., 1: 158.

Races of Man (1850). A Scottish doctor who had studied in Paris, Knox had been forced onto the lecture circuit after he was implicated in the murderous 'burking' activities of two Edinburgh body snatchers. Knox specifically targeted Prichard, accusing him of 'misdirecting the English mind as to all the great questions of race'. For Knox races were not mere 'varieties' but separate species, species that had not changed and never would.[16]

As a book *Races of Men* did not reach a large audience, and it only received a second edition in 1857. But there does seem to have been a hardening of ideas of race around 1850. We might link this to the European revolutions of 1848, in which the middle-class elites of France and various German states wrapped their demands for parliamentary democracy in racist language. A class-based, political demand was presented as the destiny of a people or *Volk*. A shared language, literature and set of customs were more than the sum of their parts; they were organs of a living national body which was now coming together. Given the organic metaphors, it was unsurprising that certain physical features and mental traits were assigned to Germans, the French, even relatively 'new' nations like the Belgians (established in 1830).[17] This process had little to do with the publication of *The Origin* and was as much about distinguishing between white European races as it was about distinguishing between white and black races. Ethnography was not a simple issue of 'us' and 'them,' whites and blacks.

The distinction between Knox and Prichard can be overdrawn. It is important to acknowledge how much room Prichard's loose language gave to would-be polygenists. The following passage comes from the concluding remarks to his discussion of West African peoples:

> On reviewing the descriptions of all the races enumerated, we may observe a relation between their physical character and their moral condition. Tribes having what is termed the Negro character in the most striking degree are the least civilized. The Papels, Bisagos, Ibos, who are in the greatest degree remarkable for deformed countenances, projecting jaws, flat foreheads, and for other Negro peculiarities, are the most savage and morally degraded of the nations hitherto described. The converse of this remark is applicable to all the most civilized races. The Fúlahs, Mandingos, and some of the Dahomeh and Inta nations have, as far as form is concerned, nearly European countenances and a corresponding configuration of the head.[18]

The reader could be excused for taking references to 'the Negro character' and 'flat foreheads' as a sign that Prichard accepted Blumenbach and

[16] Robert Knox, *The Races of Men: A Philosophical Enquiry into the Influence of Race over the Destinies of Nations*, 2nd edn (London: Henry Renshaw, 1857), p. 23.
[17] Jo Tollebeek, 'Historical Representation and the Nation-State in Romantic Belgium, 1830–1850,' *Journal of the History of Ideas* 59(2) (1999): 329–53.
[18] Prichard, *Researches*, 2: 97.

Camper's theories. The final sentence implies that European heads were differently shaped because Europeans were civilized or that the European head represented a higher form than that of the Negro.

To an extent these problems can be related to a wider looseness in the Victorians' use of 'character' (here, Prichard specifies 'physical character'; but others used it to refer to something akin to personality traits). A similar slipperiness surrounded 'civilisation'. For Smith, Kames and Prichard it was a universal human proclivity to advance through this or that stage *of* civilization. After 1850 it seemed to become something one either had or did not have. For the historian Thomas Carlyle the missionaries' hopes of civilizing freed slaves smacked of 'deep froth-oceans of "Benevolence"'. In his notorious 1849 essay for *Fraser's Magazine*, 'The Occasional Discourse on the Negro Question', Carlyle portrayed the freed slave 'Quashee' as ungrateful, feckless and lazy. It was the 'heroic' Europeans, heirs of the 'old Saxons', who had turned the jungle 'wastes' that originally clothed Jamaica into productive land. Europeans were 'wiser' than Africans and born to be masters of them.

In 1856 the bones of a Neanderthal man were discovered in a cave near Dusseldorf. Today's consensus is that the Neanderthals became extinct around 30,000 years ago, being pushed aside by later migrants from Africa (Cro-Magnons, i.e. *Homo sapiens*), despite the fact that the latter had smaller brains (one of many facts that challenge the common assumption that brains just keep getting bigger). Their classification, however, remains controverted, as some insist that there was no interaction with *Homo sapiens* (making Neanderthals a distinct species), while others point to the evidence that they bred with each other (making Neanderthals a subspecies of *Homo sapiens*). The discovery that humans, too, had a history in deep time contributed to Prichard's eclipse, insofar as it challenged the latter's view that the earth was only around 6 or 7000 years old. Deep time also provided evolutionary polygenists with a seemingly bottomless chronological pit in which to bury the common ancestor of their distinct species of human.

The study of human archaeology came late to Britain, partly because there simply weren't as many bones to be found there. That cycle of glaciation which marked the Palaeolithic period had rendered the majority of the British isles uninhabitable. Although a French archaeologist, Boucher de Perthes, had made many important finds in the Abbeville area in the 1840s, it was only with the discovery of Palaeolithic human remains in Britain, at Brixham Cave (near Kent's Cole, Devon), in 1859 that British men of science began taking evidence for pre-Deluge humans seriously. The Cave was accredited by a team delegated by the Geological Society to inspect it, and its remains were now related to Perthes' earlier findings.

Within a few years of his death in 1848, therefore, Prichard's science of 'ethnology' was looking somewhat fragile. The Ethnological Society of London (ESL) had been established in 1843 as an offshoot of the APS and partly reflected a crisis in the APS' sense of its own mission and aims. Were indigenous peoples to be studied as they were, preserved from change, or helped to 'advance' through Christianization, the introduction of Western technology and eventual 'amalgamation' with other peoples? After a failed 1841 expedition to establish a 'free' Christian community of white settlers and free slaves Hodgkin argued that such missions could not be pursued side by side with the study of ethnology. The ESL failed to establish ethnology as a serious science. It was refused its own section at the 1844 meeting of the British Association for the Advancement of Science (BAAS); it became a full section along with Geography (as Section E) in 1850, but this was mainly down to lobbying by geographers.

Prichard had not used the word 'anthropology', not because it didn't exist (it was being used by the German physiologist Karl Rudolphi in the 1820s), but because he saw it as fundamentally opposed to the 'unity of man'. In contrast to the agreeably loose definition of 'race' used by ethnologists, anthropologists had a precise, and, for Prichard, incorrect and unhelpful one:

> The instances are so many in which it is doubtful whether a particular tribe is to be considered as a distinct species, or only as a variety of some other tribe, that it has been found by naturalists convenient to have a designation applicable in either case. Hence the late introduction of the term *race* in this indefinite case. Races are properly successions of individuals propagated from any given stock; and the term should be used without any involved meaning that such a progeny or stock has always possessed a particular character. The real import of the term has often been overlooked, and the word race has been used as if it implied a distinction in the physical character of the whole series of individuals. By writers on anthropology, who adopt this term, it is often tacitly assumed that such distinctions were primordial and that their successive transmission has been unbroken.

Ever careful not to reach conclusions without data to back them up, Prichard conceded that if evidence of a 'physical character' remaining fixed from primordial times were found, 'a race so characterised would be a species in the strict meaning of the word, and it ought to be so termed'.[19]

As we have seen, Prichard was aware from the start that his views on the 'unity of man' were not shared in France and the German states. There is a similar defensiveness in the entry for 'anthropology' in the SDUK's (Society for the Diffusion of Useful Knowledge) *Penny Cyclopaedia* (1837), which sternly noted that the word 'considers man as a citizen of the world, and has nothing properly to do with the varieties of the human race'.[20] In 1856 James Hunt (1833–69) joined the ESL, becoming joint secretary in 1860. Hunt had been heavily influenced by Knox, and had no problems with the racialist baggage of 'anthropology', which he preferred to the abolitionist associations of the

---

[19]Ibid., 1: 109.

[20]Charles Knight (ed.), *The Penny Cyclopaedia of the Society for the Diffusion of Useful Knowledge*, 23 vols (London: Knight, 1834–42), 2: 97.

ESL, which he viewed in Carlylean terms, as wishy-washy sentimentality. Though he had a doctorate (purchased from a German university), Hunt was not medically trained. He practised as a speech therapist, curing stammerers (including Charles Kingsley) using a method invented by his father, a tenant farmer. In 1863 Hunt established a new group, the Anthropological Society of London (ASL), and started publishing an *Anthropological Review*. The rivalry between ASL and ESL continued until 1871, when Huxley brokered a marriage, creating the Anthropological Institute of Great Britain.

A combative character, Hunt revelled in his polygenism. His paper on 'The Negro's Place in Nature' (a play on Huxley's *Man's Place in Nature*) drew catcalls from the audience when read at the 1863 meeting of the BAAS in Newcastle. In it he argued that the 'African' could be distinguished from the 'European' not only by skin colour, but also by other physical traits, several of them ape-like (e.g. a stoop, a 'flattened' forehead), and directly related to mental and physical inferiority. The 'African' was an 'arrested development', his lesser intelligence owing to the premature fusing of the bony plates in his skull. Hunt refused to accept that mixed-race children were able to reproduce. He claimed historical evidence for his conclusion that the African could neither civilize on his own nor be civilized by contact with superior races. Though Hunt was not bold enough to refer to them as separate species, they were 'two distinct types of man'.

Published as a pamphlet the following year, it met with a very different response in the American South. As the introduction to an American edition noted, Hunt had put his finger on an 'unchanging, immovable, everlasting fact':

> We know the *fact*, and God holds us responsible only for our mode of dealing with it, and when we wilfully shut out eyes, disregard and ignore it altogether, and impiously strive to degrade our race *down*, or to force the Negro *up*, to 'impartial freedom,' or a forbidden level, we are blindly striving to reverse the natural order, and to *reform* the work of the Almighty.[21]

As the American Civil War (1861–65) raged across the Atlantic, Hunt's brand of anthropology took on added significance, as a justification for the southern states' resistance to President Lincoln, whose 1863 Emancipation Proclamation declared the end of slavery in the Confederate States of America (the rebel south). Slavery was formally outlawed by a constitutional amendment in 1865. Unsurprisingly, perhaps, the Confederate States paid a covert subsidy to Hunt's ASL, which stoutly defended Eyre's actions, to the extent of arguing that killing savages was a 'philanthropic principle'.[22]

---

[21]J. H. Van Evrie, 'Introductory,' in James Hunt (ed.), *The Negro's Place in Nature* (New York: Van Evrie, Horton and Co., 1864), p. ii.

[22]Cited in George W. Stocking, *Victorian Anthropology* (New York: Free Press, 1987), p. 251.

The ASL-ESL feud was an embarrassment to Huxley and other men of science, not so much because of Hunt's illiberal racism, but because the lack of an agreed label made it difficult to establish a discipline whose borders could be policed against cranks. After attending the Ethnological section of the 1862 BAAS meeting Kingsley wrote to Thomas Wright (ESL Secretary) that 'ethnology' was 'a refuge for the destitute for all sorts of subjects wh[ich] have not yet arrived at the dignity of sciences, and are therefore fair game for every shallow theorist'.[23] Kingsley overlooked the ESL's interest in discussing Neanderthal bones and the tools now recognized as having been left behind by the Palaeolithic inhabitants of northern Europe. This contrasted with the ASL, where Hunt's insistence on the fixity of his types left him deaf to the new science of human archaeology. When visiting speakers like Wallace mentioned Neanderthals in papers presented to the ASL in the 1860s Hunt dismissed their Palaeolithic skulls as unrepresentative, as the bones of 'idiots'.

Did the establishment of the Anthropological Institute represent a posthumous victory for Hunt (who died in 1869) or for a Prichardian 'ethnology' (now known under a different name)? On the one hand Hunt's term obviously triumphed, as did his definition of it as 'the science of the whole nature of man'. So did his use of craniometry, and belief that types could be recognized by different proportions between limbs. Both lay behind a 1870 project to collect data from indigenous peoples, which involved Huxley (as president of the Institute) signing a request circulated by the Colonial Office to all corners of the empire, asking for photographs to be taken of aborigines posing naked with arms outstretched (holding a ruler, to aid comparative measurements).

Prichard's interest in language as an ethnological tool was sidelined, or rather became philology, a distinct and isolated intellectual fief. Under the influence of the great Sanskrit scholar Max Müller, philologists were looking to an idealist German tradition. A German immigrant steeped in the German tradition of philology originating with Franz Bopp, Müller used evolutionary language, but in a curious, upside-down manner: he was endeavouring to trace all languages backwards to a perfect, ideal, God-given language, an archetype which had degenerated and become 'diseased' as it diversified into the languages spoken today.

The foundation of the Anthropological Institute was certainly a victory for Huxley's team, for the opponents of Hunt's 'fixity' interested in the origins of humans as a species, rather than classification of men as they could be found now, scattered across the globe. It was also a victory for a mood of secular respectability and detachment, something different from the ESL's fantasies of 'amalgamation' and 'improvement' on the other hand and the rather more pornographic and masochistic fantasies of inter-racial contact indulged by members of the ASL's inner circle at meetings of their 'Cannibal Club' on the other.

---

[23]Kingsley to Wright, 12 October 1862. Princeton University Library, Special Collections, AM16741.

Huxley's team included William Lubbock and Edward Burnett Tylor (1832–1917), now seen as the pioneers of British anthropology. They understood human evolution in cultural, rather than in morphological terms. 'Culture' for them embraced both the material evidence of tools and ornaments (what we would call 'material culture') as well as a set of social customs and religious beliefs. They dismissed the migrationist views of Müller and Argyll, that is, views which saw all humans as originating from an act of special creation which had been followed by a period in which 'tribes' had scattered across the globe. Under this view contemporary 'savages' such as the Tasmanians were the result of a tribe being pushed to the margins by competition and adopting 'bestial' practices as a result of the abuse of the intellectual powers given to man by God. 'Human corruption in this sense is as much a fact in the natural history of Man as that he is a Biped without feathers,' Argyll noted in his *Primeval Man* (1869).[24]

Instead of spreading out and degrading, for Lubbock and Tylor man (and, for them, human development was driven by the male) only moves upwards, climbing up a technological ladder which began with the use of whatever material lay to hand (bits of rock and wood), progressed through the fashioning of the first tools (bits of rock chipped to have one or more sharp edges) and ended up with telegraphs and steamships. Once gained, this or that advance could not be lost. Reflecting the 1860s fad for 'the comparative method', Lubbock used broad analogies between long-extinct Neanderthals and living Fuegians to construct this single ladder in his 1865 book *Pre-Historic Times*. The book included tables giving the various technologies in order, indicating how far, say, Fuegians had moved up the ladder. In its focus on gadgets and its view of moral, religious and philosophical beliefs as products of this or that 'stage of development' this ladder seemed more redolent of that 'civilisation' celebrated by H. T. Buckle in his unfinished *History of Civilization in England* (1857–61). Indeed, Tylor's first book was entitled *Researches into the Early History of Mankind and the Development of Civilization* (1865). Tylor switched to 'culture' in his most influential anthropological work, *Primitive Culture* (1871).

Lubbock cited Wallace, sharing his view that at a certain point in his evolution human morphology had become fixed, as intelligence enabled humans to use technology to make environmental adjustments (making clothing, e.g. instead of waiting many generations for hairier humans to evolve by natural selection). He was a monogenist, therefore, and was able to note how unfair it was for his contemporaries to condemn, say, a lack of marital chastity in this or that tribe when 'unchastity' was appropriate to their state of development. Yet Lubbock could, almost in the same sentence, abandon such cultural relativism for condemnation. Rather than follow Prichard in seeing fetish worship as one product of a universal

[24]Argyll, *Primeval Man: An Examination of Some Recent Speculations* (London: Strahan and Co., 1869), p. 189.

instinct which led Victorians to worship their God, Lubbock wrote that the fetish worship found across Africa could 'hardly be called a religion'. The savages found there were less intelligent than a 4-year-old (European, presumably) child. 'One of the deepest stains upon their character' was their 'cruel' treatment of women, who were viewed 'as mere domestic drudges' (they were treated as such in Lubbock's own home, Spencer might have pointed out). Lubbock nonetheless presented patriarchy as the result of millennia of progress, and even Huxley opposed admitting women to meetings of the ESL.[25]

While Tylor, Lubbock and their followers suggested that Victorians were watching their own ancestors when they observed 'savages', therefore, moral disgust fostered a tendency to see them as vile relics, rather than 'brethren' to be 'improved'. As the 1870s progressed these anthropologists could seem rather caught up in their own little world, squabbling over whether eoliths (round flints) were natural or man-made. They largely ignored Darwin's *Descent of Man*, for example. Once installed inside the new Anthropological Institute, they disappointed those, such as William Flower (1831–99), who believed that English anthropology should follow German and French anthropological institutes in classifying racial differences using craniometry and similar techniques, rather than constructing evolutionary trees for the objects used by humans, as Augustus Lane Fox did when displaying his collection of spears and boomerangs at Bethnal Green Museum in the late 1870s (and later, at the Pitt-Rivers Museum Fox founded in Oxford). Only in the 1880s, when Galton became president, did the AI engage in anthropometry.

Appointed to the Hunterian Chair of Comparative Anatomy in 1870 on Huxley's recommendation (Flower, as we have seen, had helped Huxley in the hippocampus dispute), Flower set about finding new ways of measuring skulls, using the Hunterian collections to construct a hierarchy of racial skull types with the English at the top and the Tasmanians at the bottom. A popular lecturer at the Royal Institution, Flower noted the tendency of non-European races to decline in number when they came into contact with Europeans. As the former races were (to him) unable to adapt or civilize, their extinction was inevitable. It was the job of the British Empire to collect measurements, specimens and accounts of social customs before the inevitable extinctions occurred. There was no time to be lost. A Colonial Office official wrote to Huxley noting that it had been impossible to secure a photo of a Tasmanian posing in the requested manner. 'The last man of that race died before the receipt by the Governor of Lord Granville's [i.e. the Secretary of State for the Colonies'] Despatch.'[26]

---

[25]John Lubbock, *Pre-Historic Times, as Illustrated by Ancient Remains, and the Manners and Customs of Modern Savages* (London: Williams and Norgate, 1865), p. 462.

[26]Robert Henry Meade to Huxley, 27 June 1871. Imperial College, London. Huxley Papers, Notes and Corr. (Anthropology), XV, f. 124.

Escape or extinction?

The sense of haste behind Flower's project, what has been called 'salvage ethnography', was due to his awareness of the increasing pace of European emigration to 'the colonies'.[27] Questions of empire and race were therefore closely related to Victorians' self-identification as members of a Teutonic race endowed with a migrationist instinct and the right to assimilate or exterminate any other race it encountered. This Teutonism was considered in Chapter 5, in the works of Charles Kingsley. For Kingsley empire was a curious combination of past and future, history and destiny: a way of escaping a false 'civilisation' built around cities and industrialization. For him the imperial frontier was the natural place for any Teuton, at once a return to his roots (in the German forests) and a step into an uncharted wilderness.

Kingsley marks the beginning of a surprisingly long-lasting trope of British Empire: that 'England' was to be found, not within the geographical confines of the British isles, but on the periphery. 'What should they know of England, who only England know?' asked that great imperial Englishman, Rudyard Kipling, in his poem 'The English Flag' (1891). Similar ideas are found in works by other historians of empire active in the 1870s and 1880s, such as Goldwin Smith, Edward Augustus Freeman and J. R. Seeley. Having 'followed the English' around the world in a voyage to India, the United States and Egypt, the Liberal politician Charles Dilke noted in his book *Greater Britain* (1868) that the Teutons (i.e. the English) were 'the only extirpating race'. Far from representing a fresh start in a new world, let alone a threat to Britain's primacy, the United States was just one part of 'Greater Britain'. 'Through America,' he wrote, 'England is speaking to the world.'[28]

The priority placed on racial identity made political distinctions seem irrelevant, and so Britain welcomed moves by New Zealand, Canada and Australia towards greater self-government in the closing decades of the century. Hopes of building an economic federation around this 'Greater Britain', however, proved illusory, as different dominions' economic interests pulled in different directions. As her European trading partners raised tariff barriers against her, free-trading Britain was unable to escape inside the wide boundaries of an imperial free-trade area. To make matters worse, the closing three decades of the century saw a revival of inter-European rivalry in the scramble to lay claim to hitherto unclaimed corners of Africa and

[27]James Clifford, 'On Ethnographic Allegory,' in Clifford and George Marcus (eds), *Writing Culture: The Poetics and Politics of Ethnography* (Berkeley: University of California Press, 1985), pp. 99–121 (112).

[28]Cited in Theodore Koditschek, *Liberalism, Imperialism and the Historical Imagination: Nineteenth-century Visions of a Greater Britain* (Cambridge: Cambridge University Press, 2011), p. 207.

Study cultures

Greater Britain

the South Seas. Even that high priest of small government and balanced budgets, Gladstone found it difficult to resist demands that Britain match the efforts of France and now newly unified Germany to plant the flag in areas of little economic interest, areas that were never likely to appeal to would-be emigrants.

The 'Greater Britain' and 'imperial preference' camps found in empire an opportunity for 'the coming race' (i.e. the Teutons) to escape the paradoxes of 'civilisation' evident at home in the shape of unemployment, slums, union unrest and the rise of socialism. From the 1890s onwards, however, others drew a very different picture of 'the coming race', confronting Britain, her empire and her civilization with a range of threats, which were no less frightening for being vague and sometimes contradictory. Degradation, immigration and the 'taint' of heredity associated with the discoveries of Weismann and Mendel have already been mentioned. But Victorian science fiction writers found many more: genetic modification, airplanes, electricity, killer smog and Egyptian mummies.

The Coming Race (1871) by Edward Bulwer-Lytton was the first of what would be a series of fantasies in which male engineers, zoologists or men of science stumbled across a race of superior human-like creatures, only for these attempts to 'conquer' unknown territories back-firing (when the 'discovered' decide to 'conquer' back). A shopkeeper's son, H. G. Wells (1866–1946) was apprenticed to a draper's before he managed to persuade his family to let him train as a teacher, spending a year under Huxley's tutelage at the Normal School of Science.

Wells' dystopian fantasy novel *The Time Machine* was published in 1896, and depicted a future in which mankind has evolved into two species: one a gentle, happy race of idlers (the Eloi, who live on the surface of the earth), the other a horde of bestial if industrious Morlocks (who toil underground, but who venture up at night to prey on their happier cousins). A Fabian socialist, Wells depicts the long-term consequences of industrial capitalism, with its creation (at least, according to socialists) of a class of investors whose life of luxury is maintained by proletariat hordes whose existence is entirely foreign to them. Other fantasies of the 1890s depicted 'modern' societies consumed by the ever-expanding gulf between technological progress (which endows humans with unprecedented powers), and humanity's baser instincts (which condemn humans to abuse these powers in fantastically destructive ways). Such fantasies reflected broader public disenchantment with science. Rather than enhancing individual hopes and freedoms, scientific 'progress' seemed to be fostering alienation, bureaucratization and an impersonal mechanization. To us these are all recognizable symptoms of modernity. But this was not the future Victorians felt they had signed up for.[29]

[29]See Roy M. MacLeod, 'The "Bankruptcy of Science" Debate: The "Creed of Science" and its Critics," ' in MacLeod (ed.), *The 'Creed of Science' in Victorian England* (Aldershot: Ashgate, 2000), pp. 1–22.

Whether these 'savage' bestial urges were located within otherwise 'civilized' Victorians at the heart of empire or assigned to 'lower' races invading from 'the beyond' (Prussians, Slavs, Orientals and extra-terrestrials), the assault on imperial assumptions was clear. Were the Victorians in fact 'the coming race'? Did technological progress always advance in step with 'civilisation'? Were Britons really secure, really 'at home' on the imperial periphery, or was that precisely the place where their 'civilisation' was most exposed? Even as Colonial Secretary Joseph Chamberlain made Queen Victoria's 1897 jubilee into a spectacular celebration of 'Greater Britain', chinks in the imperial armour were appearing.

These chinks were exposed just 2 years later, when the second Boer War erupted in South Africa. The success with which small bands of Dutch settlers (or 'Boers') from the Transvaal Republic and Orange Free State kept regular forces drawn from across the British Empire at bay astonished Britons at home. So did the policies of burning Boer homesteads and interning Boer families in camps. Army Medical Council reports suggested that almost half of would-be recruits were physically unfit to serve their country. Combine these developments with the Prussian emperor's glee at Boer successes and the picture became even more confusing. Britain signed an alliance with France in 1904 and fought with her against Germany 10 years later. The war seemed to expose something sick at the heart of the imperial project, setting 'Teuton' against 'Teuton' (Britain against Prussia) and calling Birchenough's strong, unified 'Greater Britain' into question as never before.

Further reading

Banton, Michael, *Racial Theories* (Cambridge: Cambridge University Press, 1998).

Bell, Duncan, *The Idea of Greater Britain. Empire and the Future World Order, 1860–1900* (Princeton: Princeton University Press, 2007).

Brantlinger, Patrick, *Taming Cannibals. Race and the Victorians* (Ithaca: Cornell University Press, 2011).

Burrow, John, 'Evolution and Anthropology in the 1860s: The Anthropological Society of London,' *Victorian Studies* 7(2) (1963): 137–54.

—, 'The Clue to the Maze: The Appeal of the Comparative Method,' in John Burrow, Stefan Collini and Donald Winch (eds), *That Noble Science of Politics. A Study in Nineteenth-century Intellectual History* (Cambridge: Cambridge University Press, 1983), pp. 207–46.

Fee, Elizabeth, 'The Sexual Politics of Victorian Social Anthropology,' in Mary S. Hartman and Lois Banner (eds), *Clio's Consciousness Raised: New Perspectives on the History of Women* (New York: Octagon, 1979), pp. 86–102.

Hall, Catherine, *Civilising Subjects. Metropole and Colony in the English Imagination, 1830–1867* (Cambridge: Polity, 2002).

Koditschek, Theodore, *Liberalism, Imperialism and the Historical Imagination. Nineteenth-century Visions of a Greater Britain* (Cambridge: Cambridge University Press, 2011).

Lloyd, Trevor, *Empire. The History of the British Empire* (London: Hambledon and London, 2001).

McNabb, John, *Dissent with Modification. Human Origins, Palaeolithic Archaeology and Evolutionary Anthropology in Britain, 1859–1901* (Oxford: Archaeopress, 2012).

Richards, Evelleen, 'Huxley and Woman's Place in Science: The "Woman Question" and the Control of Victorian Anthropology,' in James R. Moore (ed.), *History, Humanity, and Evolution: Essays for John C. Greene* (Cambridge: Cambridge University Press, 1989), pp. 253–84.

Stepan, Nancy, *The Idea of Race in Science: Great Britain, 1800–1960* (London: Macmillan, 1982).

Stocking, George W., *Victorian Anthropology* (New York: Free Press, 1987).

CHAPTER SEVEN

Progressive evolution? Herbert Spencer, social science and 'Social Darwinism'

FIGURE 4 *Unknown Photographer,* Herbert Spencer, *c. 1898. Caricaturists took great delight in Spencer's domed head, which seemed fitting in the author of the Synthetic Philosophy. This albumen cabinet print by contrast hints at the sitter's more playful side: the merry, darting eyes in the otherwise owlish face.*

'No society has ever been more committed to progress as a central notion or goal than Victorian Britain at the height of its colonial and industrial expansion.' So wrote the evolutionary biologist and historian of science Stephen Jay Gould in 'A Tale of Two Work Sites', an essay published in 1997. In it Gould reflected on the history of his office at New York University. In 1911 a sweatshop had stood on the same site, one in which recently arrived European immigrants assembled ladies' blouses. One Saturday a fire broke out, which ended up killing 146 workers, most of them women. Although the cause of the fire was unknown (overheating machinery and an illicit cigarette butt have been proposed), the death toll was widely blamed at the time on insufficient building safety regulations and on the owners having locked the fire exit to prevent pilfering by their employees.

It was difficult to pin down who exactly was to blame, Gould conceded. But it was clear that the guilty party had an accomplice, someone who provided them with intellectual justification for their ruthless pursuit of profit. The accomplice was the Victorian evolutionary thinker Herbert Spencer, and the justification was Social Darwinism: a philosophy then in favour among leading American industrialists – and now, Gould noted, among 'our 'modern' ultra-conservatives'. Spencer had constructed an analogy between society and a living organism, understanding both as a delicate arrangement of interdependent organs. He had fiercely opposed any attempt by the state to meddle with this delicate arrangement, advocating a strict *laissez-faire* attitude towards economic injustice, opposing public education and welfare (including health and safety regulations) as well as trade unions. Free-market capitalism was an extension of natural selection, enriching the winners and starving the losers.

Darwin had, Gould conceded, adopted Spencer's phrase 'the survival of the fittest' (first used in 1853) in later editions of *The Origin*. Unlike Spencer, however, Darwin did not see understanding of natural selection as qualifying him or anyone else from seeking to apply any supposed 'lessons' for human society. The fire showed the 'palpable influence of a doctrine that applied too much of the wrong version of Darwinism to human history'. 'I do not doubt that the central thrust of Social Darwinism – the argument that governmental regulation can only forestall a necessary and natural process – exerted a major impact in slowing the passage of laws that almost everyone today, even our archconservatives, regard as beneficial and humane.'[1] Gould's view of Spencer is one widely shared among historians of science, who also find Spencer a useful lightning rod, someone on whom to deflect criticism that might otherwise hit Darwin.

Others wonder if it is helpful to speak of Social Darwinism at all. The term was not widely used before the publication of the American historian Richard Hofstadter's doctoral dissertation on *Social Darwinism in*

[1]Stephen Jay Gould, 'A Tale of Two Work Sites,' in Stephen Jay Gould (ed.), *The Lying Stones of Marrakech: Penultimate Reflections in Natural History* (New York: Vintage, 2001), pp. 251–68 (265).

American Thought (1944). A former Communist, Hofstadter's work was inevitably coloured by his political views, while his reading of Spencer was uninformed by archival research. More recently historians of science have argued that there is no difference between Darwinism and Social Darwinism. Darwin's ideas were built on speculations concerning relationships between populations (not just individuals), and so all Darwinism is social. Interest in Spencer's writings remains restricted to historians of political thought, however. The punch of his political writings, published as *Man vs. The State* (1884), may have encouraged a tendency to consider Spencer solely as a political thinker, as an extreme libertarian who (as Peter Bowler claims) saw *laissez-faire* capitalism as the highest form of society.[2]

An uncharacteristically fierce set of polemics, it will always be something of a challenge to present Spencer as a serious thinker on evolution as long as *Man vs. The State* remains the only one of his works available in a modern scholarly edition. Yet that is what this chapter will attempt to do. In view of current debates over the term, it will deliberately avoid using the term 'Social Darwinism'. As we shall see, Gould's account of Spencer's thought is limited. Gould on the one hand presents a government or state as a 'given' (a view Spencer did not share) and links Spencer with a High Victorian 'Gospel of Work' (which Spencer ridiculed). The term risks reducing the overarching 'question of questions' (to use Huxley's phrase) – that is, whether humans can, or ought to, construct an ethics independent of our inherited instincts for self-preservation – into a question of politics, rather than a tool for thinking with. The reader is encouraged to reach his or her own conclusion on the value or otherwise of 'Social Darwinism' (as a term), after having read both this chapter and Chapter 9, which addresses the Darwinian Left.

The intellectual power and coherence of Spencer's thought remains suspect largely because of the ambition and scope of his Synthetic Philosophy, which sought to establish the sciences of biology, sociology, psychology, even morality on the basis of a set of universal laws of development. Who today would even imagine something so wide ranging? Spencer's project is condemned on the grounds of its supposed arrogance. Yet the span of Spencer's thought (and even, perhaps, his personal eccentricities) only made his ideas more fascinating to Victorian audiences. Spencer found evidence of 'evolution' everywhere, from the formation of galaxies to schoolyard gangs and the banal salutations we use every day. He offered an internally consistent view of the world that appealed to Liberal Christians. For others it came to replace the Christian view in its entirety.

Writing and reading the later volumes outlining the Synthetic Philosophy became something of a chore for Spencer and his readers after 1880, one suspects. By then Spencer's authority was beyond question. He was a leading public intellectual at a time when such figures were lionized

[2]Peter J. Bowler, *Evolution: The History of an Idea* (Berkeley, CA: University of California Press, 1984), p. 99.

more than ever before or since: courted by prime ministers, offered safe parliamentary seats and invited on globetrotting lecture tours. To Victorians Spencer *was* 'evolution', in the same way that Stephen Hawking *is* the Big Bang and Black Holes. As with Hawking's *Brief History of Time* (1988), few of Spencer's many admirers actually read the books they bought, but they were exhilarated by the vistas he opened up. Even if they didn't quite follow his meaning, Spencer's sonorous phrases reassured Victorians that the riddles surrounding them had an answer. Progress was 'a beneficent necessity'. Evolution was happening everywhere, not just in buried rocks and distant finches' beaks, but here and now, to you and me. If Kingsley emphasized effort and activity, Spencer gave that energy a direction and a name.

Springs of action: Childhood and youth

The son of a Derby schoolmaster, Herbert Spencer grew up in an environment in which discussion of politics, theology and the natural sciences was encouraged. His father George served as secretary to the city's Philosophical Society, an institution proud to count Erasmus Darwin and Joseph Priestley among its former members. Although Herbert's uncle was a Cambridge-educated Anglican priest, he too exhibited a proud provincial scepticism of established authorities and a stolid faith in common sense. Only Herbert's mother, a plumber's daughter, was left out. In a family where opinions were confidently advanced and stoutly defended, Harriet Spencer's quiet, fussy nature was at a severe disadvantage, to the extent of her being victimized by her husband, a situation which worsened after he lost capital invested in an ill-fated lace manufactory.

George Spencer was an innovator in more ways than one, albeit one with a tendency to lose himself in endless fine-tuning. As a teacher he dispensed with the usual rote learning in favour of a more Socratic, student-led approach, addressing subjects far beyond the reach of the traditional, classics-dominated curriculum of the day. Though he was educated at home, the eldest of a brood of siblings (none of whom survived to adolescence), Herbert seems to have been left to his own devices for long periods, told to teach himself. When in 1833 he was sent to his uncle Thomas's rectory in Somerset to study he initially rebelled against the new, stricter regime.

Thomas Spencer was making a name for himself as an outspoken advocate of free trade, temperance and the New Poor Law. Herbert's first publication was a letter to the *Bath Magazine* challenging an opponent of his uncle's hard-headed approach, which had led to a dramatic decrease in the Poor Rate paid by his parishioners, from £700 a year to just £200. The old Poor Laws had not served to advance 'civilisation', the 16-year-old Herbert insisted, but only to foster 'ignorance, drunkenness, and crime' among an idle 'set of villains'.[3] Though precocious in many ways, Herbert's character was also marked by a certain laziness. Unable to keep to a course of classical reading intended to prepare him for study at Cambridge, in 1836 Herbert was sent to work for an engineer. He spent several happy years designing stations and surveying routes for various Birmingham-based railway companies.

As already noted, the arrival of the railways as well as the telegraphs, which ran along them, changed Britain profoundly and may have inspired Spencer to reflect on how they quickened the 'pulse' of the nation, changing 'actions, thoughts, emotions'.[4] An unsuccessful attempt to woo his boss's flirtatious (and already committed) niece on the other hand had profound personal repercussions. Though Spencer would in turn be courted by Marian Evans (better known under her pen name, George Eliot), he never again contemplated marriage or romantic attachment of any kind. Imprinted by the experience of his parents' unhappy marriage and seven deceased siblings, Spencer seems to have identified marriage and reproduction with violence, domination and death. In the same 1853 essay in which he coined the phrase 'survival of the fittest' Spencer propounded a law by which the fertility of organisms declined as they became more highly developed.

Historians should be careful of explaining any thinker's ideas solely in terms of upbringing. Herbert's early life is nonetheless significant because his philosophy crystallized so early and thereafter remained largely unchanged, partly thanks to his lazy refusal to read other philosophers, even Kant. The 12 letters Spencer wrote to *The Nonconformist*, a newspaper, on 'The Proper Sphere of Government' (published separately in 1843) set the terms of his thinking for the next 61 years. In them Spencer defined the state's role as maintaining justice, its duty 'to protect person and property'. Anything else would constitute foolish, presumptuous meddling that would upset 'that beautiful self-adjusting principle' by which all the different elements of Creation were regulated.[5] The state should therefore stay out of religious matters and disestablish the Church of England. It should stop intervening in the grain markets, and hence abolish the Corn Laws, which kept grain prices artificially high. It should give up the empire and disarm unilaterally.

Much of this echoed Adam Smith. Whereas Smith had emphasized the mysterious nature of the 'self-adjusting principle' (his 'invisible hand'), however, Spencer implied that the principle might be investigated and understood. There was a whiff of George Combe and phrenology, too,

[3]The letter is reproduced as an appendix to John Offer (ed.), *Herbert Spencer. Political Writings* (Cambridge Texts in the History of Political Thought, Cambridge: Cambridge University Press, 1993), p. 180.

[4]Spencer, 'Progress: Its Law and Cause,' in Spencer, *Essays: Scientific, Political, and Speculative*, 3 vols (London: Williams and Norgate, 1891), 1: 8–62 (58).

[5]Offer (ed.), *Political Writings*, p. 6.

in 'The Proper Sphere', in so far as its climax presented well-intentioned government measures as robbing Englishmen of the opportunity to flex organs of forethought and intellect:

> Establish a poor law to render his forethought and self-denial unnecessary—enact a system of national education to take the care of his children off his hands—set up a national church to look after his religious wants—make laws for the preservation of his health, that he may have less occasion to look after it himself—do all this, and he may then, to a great extent, dispense with the faculties that the Almighty has given to him. Every powerful spring of action is destroyed—acuteness of intellect is not wanted—force of moral feeling is never called for—the higher powers of his mind are deprived of their natural exercise, and a gradual deterioration of character must ensue. Take away the demand for exertion, and you will ensure inactivity. Induce inactivity, and you will soon have degradation.

It was not cruel, Spencer insisted, to stand back and let the incapable die. 'Many arrangements in the animal creation cause much suffering and death,' he noted, 'but we do not thence infer that the Almighty is unmerciful.'[6]

Having failed at his first attempt to make it as a journalist in London, Spencer returned to the railways in 1845. Like his father he tinkered with a number of inventions, hoping one or other would free him from the need to work. With his uncle's help he secured a job as sub-editor of *The Economist*, the leading free-trade newspaper, still in existence today. The job gave him a way into the leading intellectual circles in London, most importantly that of G. H. Lewes. Spencer approached Huxley in 1852 after reading the latter's article on hydrozoa, beginning an important friendship. In 1853 the first signs of ill health emerged. Eventually this unidentified nervous complaint would imprison Spencer, both mentally and physically.

Statics and kinetics

In 1851 Spencer published his first book. Unlike *The Proper Sphere of Government*, which he had published at his own expense (losing a considerable amount of money), this time his uncle had persuaded a publisher, John Chapman, to take the risk. Where *The Proper Sphere* was a polemic pamphlet laced with references to current affairs, *Social Statics* was Spencer's attempt to construct a 'true social philosophy'. The title seems odd to us. What are 'statics'? Here, as in many other cases, it is useful to consult the *Oxford English Dictionary* (*OED*) whose definitions not only assist in teasing apart the different senses of a word, but are also accompanied with

---

[6]Ibid., p. 48.

short quotations illustrating how the word was used at different points in time. The *OED* defines 'statics' as 'the branch of physical science concerned with the action of forces in producing equilibrium or relative rest, and with the forces acting on systems and structures in equilibrium.' It quotes Whewell using it in 1847 to refer to a division of 'mechanics', along with 'kinetics'.

There is a whiff of Spencer the railway engineer, therefore, in his choice of title. Spencer's 'social philosophy' understood humanity as a dynamic entity, as a collection of organs, instincts and passions which needed to be kept in balance. Combe's term 'constitution' (as in *The Constitution of Man*) has a similar sense, and like Combe, Spencer defined happiness as the exercise of all one's faculties, rather than the overuse of this or that faculty in isolation. Statics might remind us of 'stasis', of immobility. Once equilibrium had been achieved within an individual, maintaining that equilibrium would be easy. One would think. Spencer was aware that the environment surrounding the individual was constantly changing. The equilibrium was a moving target, therefore; it was constantly readjusting to 'fit' new conditions. The universe or God (Spencer was evasive here) had so shaped 'the human faculties' that they propelled humans into a group, into 'the social state'. In obedience to 'a law underlying the whole organic creation', humans *had* to 'be moulded into complete fitness for the social state; so surely must the things we call evil and immorality disappear; so surely must man become perfect'.[7]

Yet Spencer's 'social state' was as fragile as it was relentless. Spencer often chose to represent it as a victim to 'false' social philosophy, to utilitarianism, a philosophy oriented towards 'the greatest happiness of the greatest number'. Spencer described this as the 'expediency principle', which he saw as wedded to 'the eternity of government'. 'It is a mistake to assume that government must necessarily last forever,' Spencer wrote, claiming that the government which utilitarians relied on (to carry out this or that policy) was not an essential part of human society. Though necessary in one stage of 'civilization', in one phase of mankind's progressive social development, it was not a universal 'given'. Indeed, it would have to pass away if progress was to continue.[8]

It is difficult to exaggerate just how revolutionary this view was, in terms of the history of political thought. From the Stoics of ancient Greece through Enlightenment figures like John Locke and Jean-Jacques Rousseau to contemporaries of Spencer like the great J. S. Mill, the individual had been understood as passing from a lawless 'state of nature' into a community of law. This was often understood in terms of a contract: certain freedoms or 'natural rights' were sacrificed in exchange for other freedoms or for protection against other individuals. Although some philosophers romanticized the 'state of nature' and others questioned the terms of 'the

[7]Ibid., p. 60.

[8]Herbert Spencer, *Social Statics* (New York: Robert Schalkenbach Foundation, 1970), p. 13.

social contract', very few were willing to join Spencer in his refusal to see 'government' as fundamental to human society. His colleague at *The Economist*, Thomas Hodgskin (1787–1869), was one important exception. Looking further afield, one can see similarities with the thought of the Russian anarchist Mikhail Bakunin (1814–76).

For Spencer 'government' was an assault on human nature from 'outside', a force which a powerful minority used to oppress fellow humans. One might see the distinction between 'statics' and 'kinetics' as helpful here. Theories of government like utilitarianism, party politics and the suffrage question were not 'social statics', but 'social kinetics'. The *OED* defines 'kinetics' as 'the branch of dynamics which investigates the relations between the motions of bodies and the forces acting upon them,' in opposition to 'statics', which 'treats of bodies in equilibrium'. Like 'government', 'kinetics' assumes a distinction between force and impulse, 'doer' and 'victim'. It can also suggest that there is a single cause and a single effect. Spencer had seen enough of what agents could do to others in his own family. It was no different from the violence unleashed by war, or party political conflict. Not only did it cause immediate harm, but it also encouraged others to flex their 'lower' faculties. Not only aggression, in this case, but also submission, an acceptance that 'might is right'. Violence and submission may have helped kick-start civilization, holding primitive individuals together in small groups, but the 'social state' that lay ahead was non-violent. There was no place in it for the kind of 'hero-worship' of Victorian historians like Carlyle (and Kingsley, too).

'Kinetics' weren't just dysgenic and morally wrong, they were bad science, too. 'Kinetics' presume that a certain amount of energy is created and then transferred from one thing to another one thing. 'Statics' recognizes that the amount of energy in the universe is fixed: all we can do is convert it from one form to another, a process which is never 100 per cent efficient. We inhabit a massive, yet closed system, one whose equilibrium is moving. 'It is impossible for man to create force,' Spencer observes, and that goes for 'moral force,' too. Forcing individuals to contribute money to support the poor (in the form of taxes) will not make those individuals more charitable; on the contrary, they will pay grudgingly, and may come to resent the poor.[9] Spencer's disdain for concepts of reason, will, democracy and representation may seem odd for a Victorian intellectual. In its attention to 'the science of energy', however, *Social Statics* was cutting edge.[10]

Spencer's tendency to draw analogies between our individual passions and abstract forces familiar from physics can feel somewhat cold, unfeeling and deterministic. Something in us rebels against being described as an 'atom' repelled or attracted by this or that force or passion. We prefer to

[9]Ibid., p. 240.
[10]Crosbie Smith, *The Science of Energy. A Cultural History of Energy Physics* (Chicago: Chicago University Press, 1998).

be called 'citizens'. In doing so, Spencer reminds us, we subject ourselves to a 'God upon earth' whose shoots were watered by bloody oppression and enslavement. We celebrate our enslavement to a false idol, one to which we foolishly attribute great power, even as it continues to disappoint us, over and over again.[11] Even when the state attempts to provide something quite basic such as legal redress, it fails: the rich enjoy better lawyers, while the poor are largely excluded. This view may seem extreme to us, but the questions it asks remain challenging.

It is important to bear these underlying principles in mind, as otherwise it is easy for the reader of *Social Statics* to be distracted by the specific policies Spencer recommends. Many of these go against the ghoulish (and Gouldian) caricature of Spencer. All involve Spencer deliberately seeking to take us back to the demands of 'equity' and 'justice' by removing the blinders imposed by centuries of custom, itself the product of violent history. 'Pull to pieces a man's Theory of Things,' Spencer noted, 'and you will find it based upon facts collected at the suggestion of his desires.' Thus greed can disguise itself as a virtue. 'Among money-hunting people a man is commended in proportion to the number of hours he spends in business; in our day the rage for accumulation has apotheosized work; and even the miser is not without a code of morals by which to defend his parsimony.'[12] So much for Spencer as a champion of big business!

Spencer asserts that the land belongs to all the people, and should be returned to them, then rented out to whomever wished to till the land (to the previous owners, if they wish). All colonies should be given back to the indigenous peoples who have been cruelly oppressed by colonial powers for far too long. Children have the same rights as adults (including the vote) and should not be beaten. Women should be enfranchised, too. Spencer's coruscating prose spears Eurocentric imperialism and misogyny in one paragraph:

> That a people's condition may be judged by the treatment which women receive under it is a remark that has become almost trite . . . Yet strangely enough, almost all of us who let fall this observation overlook its application to ourselves. Here we sit over our tea tables and pass criticisms upon national character or philosophize upon the development of civilized institutions, quietly taking it for granted that we *are* civilized – that the state of things we live under is the right one, or thereabouts. Although the people of every past age have thought the like and have been uniformly mistaken.[13]

Extraordinary, unbelievable words coming from a man, a British man, living in the capital of the largest empire the world had ever seen, in

[11]Spencer, *Social Statics*, p. 186.
[12]Ibid., pp. 141–2.
[13]Ibid., pp. 143–4.

the year of the Great Exhibition, when the Victorians' faith that they represented the highest stage of 'civilization' and 'progress' was stronger than ever before.

The laws of development

Encouraged by reading Von Baer's embryology in 1851 and his friendship with Huxley, Spencer followed *Social Statics* with a series of essays in which he gave his ideas a more biological dress. In 'Progress: its law and cause' (1857), originally published in the radical *Westminster Review*, he advanced one of his laws of development, according to which everything evolved 'from the homogenous to the heterogenous'. Whatever span of time or space one chose, whatever order of magnitude one selected, the same pattern was discernable. Evolution had fractal symmetry. Embryological recapitulation showed the same process happening on three different time frames within a single organism. Whether one considered a cloud of interstellar dust coalescing and beginning to spin (the nebular hypothesis) or a fertilized egg beginning to divide, everywhere the diffuse, simple, unstructured and isolated was becoming condensed, complex, intricate and specialized.

'The Social Organism' (*Westminster Review*, January 1860) constructed a detailed analogy between the 'social state' of humanity and other forms of life. Just as higher forms of life had more varied and specialized organs, so more developed civilizations had distinct professions. Whereas Bushmen society was analogous to a simple protozoan, British society was like a human body, an analogy which saw Spencer relate the bicameral parliament at Westminster to the two halves of the human brain and the platelets carried by the blood to money circulating the economy.[14] The only place where this analogy broke down was in the case of will: the social organism did not have a collective 'mind' and will to match the single 'mind' and will of the human being.

In his 1857 essay Spencer asked his readers to imagine a group of the same species divided by continental uplift, that is, by the formation of a mountain range. Climate and other factors would make the environments on either side different, which would in turn cause variation 'in different species, and also in different members of the same species'.[15] New species would result. He also asked them to consider all the many species of dog which had resulted from selective breeding by humans. In his earlier works he had noted how competition among individuals led to specialization. His strong emphasis on the influence of the environment, however, led Spencer to see such competition, not as a means of speciation, but as a 'purification'

[14]Spencer, 'The Social Organism,' in Spencer (ed.), *Essays: Scientific, Political, and Speculative*, 1: 265–307 (302, 293).

[15]Spencer, 'Progress: Its Law and Cause', in Ibid., 1: 49–52.

exercise, a means by which a species was kept 'up to the mark' in terms of its adaptive relationship to its environment. Spencer certainly stressed the 'malleability' of life forms, but this was not the same as 'variability': in the former model life forms were relatively passive, in the latter they are constantly mutating, producing random changes, which are then selected. Never having had much call to note 'the species problem', Spencer never showed much interest in how speciation occurred. He did not, in other words, come up with the theory of natural selection before 1859.

In 1859 Spencer was preoccupied with working out how to support himself while he formulated his Synthetic Philosophy. He had drawn up a prospectus in which the contents of each of the ten volumes were sketched out, following that flash of insight which had come to him the previous year. Although he had inherited a significant sum on his uncle Thomas's death, Herbert had run through it, travelling around Europe in search of stimulation as well as treatment for his unidentified nervous malady. With the help of the American publisher Edward Youmans Spencer was able to secure enough subscriptions to his unwritten volumes for him to get started. When this support dried up in 1866, a second push by Youmans and J. S. Mill got the Philosophy back on track. By the 1870s royalties on the highly popular early volumes were bringing in a healthy income.

First Principles (1862) laid out the basic structure and epistemological basis of the Synthetic Philosophy. In the first part Spencer explained that the struggle between science and religion did not represent an inevitable stage in human development. It was simply the result of a failure to draw the border between the two in the right place and in the right way. The job of science was to study 'the knowable', while that of religion was to investigate the 'unknowable', as well as to consider the status of the matter addressed by science, the bases of science's claim to objective knowledge. There was, Spencer declared, an absolute reality (what we might call a 'ground of being') where existence was not 'conditioned', was not subject to change dependent on the observer's position relative to it. There was an 'absolute Cause', a God-like Creative agency which could not, however, be influenced by prayer. Drawing on the philosophers William Hamilton and in particular Hamilton's student, the High Churchman Henry Mansel's 1858 Bampton Lectures, this was a safely British 'philosophy of the unconditioned'. It helped Spencer to defend himself from accusations that he was following the positivist model of the French philosopher Auguste Comte, whose equally wide-ranging *Course in Positive Philosophy* (1830–42) had entirely detached the sciences from metaphysics.

Part Two of *First Principles* was devoted to 'the Laws of the Knowable', such as the laws of physics (what we know as the First and Second Law of Thermodynamics) and laws of 'evolution'. Here evolution 'from the homogenous to the heterogenous' took centre stage, along with other processes which are equally abstract, such as the 'advance in definiteness' or 'individuation'. Rather than staying with a specific example drawn from, say, comparative anatomy, Spencer jumps from annelid worms to etymology to Assyrian wall sculptures to child development, sometimes within a page or two – hopping from one illustration of a particular process to another. Darwin is hardly mentioned at all, and then only fleetingly. Spencer notes that the forms of amoebas (single-celled organisms) are so variable and 'indeterminate' that 'many forms which were once classed as distinct species . . . are found to be merely varieties of one species'. Meanwhile the 'highest organisms' are far more 'precise in their attributes', which remain constant under 'changed conditions'.

> If, however, species and genera and orders have resulted from the process of "Natural selection," then, as Mr Darwin shows, there must have been a tendency to divergence, causing the contrasts between groups to become more and more pronounced. By the disappearance of intermediate forms, less fitted for special spheres of existence than the extreme forms they connected, the difference between the extreme forms must be rendered more decided; and so, from indistinct and unstable varieties, must slowly be produced distinct and stable species.[16]

Evolution here is Darwinian in a sense, but seems more like a process whereby a blurry cloud of diverse life comes into focus. Natural selection is creating a tree, but a topiary one. Instead of sprouting branches and twigs from a single trunk, it is clipping away at the bush of life until it resembles a Darwinian tree.

Aided by a series of research assistants, Spencer collected examples which could be threaded together in the form of yet more analogies that showed 'laws' at work in the universe. The two volumes of the *Principles of Biology* appeared in 1864 and 1867. Here again Spencer's delight in analogies is such that it can be hard to pin down what exactly is evolving, what the unit of selection is. Sometimes it is one nation-state competing with another nation-state, sometimes one tribe with another, one species with another – or even one organ with another organ, within a single organism. 'The network' or simply 'life' itself seems to be the hero of his story of evolution, of the organization of life (and life as organization).

Man Versus the State

Having covered biology and psychology, Spencer's 40-year campaign to complete the Synthetic Philosophy reached the science of human society, 'sociology' (Spencer had borrowed the word from Comte). *The Principles of Sociology* was published in three volumes between 1876 and 1897. It

[16] Spencer, *First Principles* (London: Williams and Norgate, 1862), pp. 186–7.

embraced humanity's aesthetic, religious, linguistic and industrial evolution. Spencer traced the evolution of religious belief from ancestor worship to polytheism to monotheism, for example. He also traced humanity's evolution from militarism to industrialism, seeing the two as distinct stages. Spencer's assumptions in this regard were those of a Free Trader of the 1840s and 1850s. Although the term 'military-industrial complex' would not be coined until the following century, the synergy between private industry and the state was already pretty obvious. Spencer need only have travelled to London's Hatton Garden to meet Hiram Maxim, inventor of the single-barrelled machine gun and new varieties of high explosives, who profited mightily from the many low-level conflicts fostered by British imperial expansion.

Continuing his theme of evolution as 'individuation', *The Principles of Sociology* predicted that humanity's freedom to flex and perfect the full range of his faculties would increase. Unless, of course, government's appetite for ever-greater intervention, regulation, inspection and 'improvement' straightjacketed the social organism, throwing its upward development into reverse, towards slavery rather than towards freedom. The state was breaking that 'law of equal freedom' which Spencer had advanced in *Social Statics*, that 'each has freedom to do all that he wills provided that he infringes not the equal freedom of any other.' The book we know as *Man Versus the State* (1884) originated as a series of articles in the *Contemporary Review*. In the four decades that had passed since *The Proper Sphere* Spencer's position had not changed.

His audience had. Thanks to the fame of his Synthetic Philosophy, it was far, far larger. It was also far more accustomed to state intervention in all areas of life. Indeed, many readers would have pointed to this expansion of the state as evidence of social evolution. Though 'economy' remained a popular watchword, the concerns of venality ('jobs for the boys') which Spencer had expressed in 1842 had been addressed by reforms to the civil service introduced in the wake of the Northcote-Trevelyan Report (1854). With admission to the service now determined by the results of a competitive examination it was hard to characterize state bureaucrats as a bunch of self-serving second-sons of the aristocracy. Spencer's attack on the mission creep of what he confusingly called 'the new Toryism' could seem outdated, therefore.

Whereas state intervention in education had been highly controversial when first seriously proposed in the 1830s, by 1884 compulsory state education up to the age of 12 was the norm. The General Board of Health established in 1848 after a series of deadly outbreaks of cholera had originally few powers and needed legislative confirmation on an annual basis. In 1853 Parliament passed an Act making vaccination compulsory. By 1884 public health and sanitary science were professions and disciplines in their own right, as well as parts of the government machine. Those parents and guardians who refused to have their children vaccinated were made to pay a fine. As Huxley noted in an 1871 critique of Spencer's

'Administrative Nihilism', an individualist insistence that the state act merely as 'policeman' enforcing the law of equal freedom overlooked the fact that individual A could find his freedom harmed if his neighbour B exercised *his* freedom to, say, leave his drains unblocked (causing cholera which might infect A as well as B) or leave his children uneducated (causing crime whose costs were also borne by A). The failure of state intervention to which Spencer pointed would, Huxley promised, be remedied as legislators learnt from experts.

THE SOCIAL SCIENCE ASSOCIATION

Established in 1857, the Social Science Association (SSA) was one forum in which Britain's ruling elite could learn from experts, in the public gaze rather than within Whitehall (the area of London which housed government offices) or the Houses of Parliament (e.g. in the course of select committee investigations). Like the BAAS the SSA's conference convened in a different city each time and drew large audiences, at least, up until 1879. This middle-class audience was attracted by the opportunity to hear leading politicians, philosophers and men of science and industry speak on topics of the day. Unlike the BAAS women were allowed to be full members and even give papers. The SSA served as a 'think-tank' (to use a twentieth-century term), heavily influencing legislation including the Habitual Criminals Act (1869), Married Women's Property Act (1870) and the Prevention of Crimes Act (1871).

For a body with Liberal sympathies, however, these successes were double-edged. The very title of the Habitual Criminals Act implied that there was a certain class of people who were predisposed to offend. On the 'prevention better than cure' principle the SSA urged the police to treat these people differently, regardless of whether they had yet to commit a crime. The age-old principle of the presumption of innocence was under threat. The Italian criminologist and self-proclaimed Darwinian Cesare Lombroso (1835–1909) identified physical traits that marked out 'born criminals', traits that represented 'reversion' to an earlier, savage stage of development.

The SSA owed its origins to a strategic partnership between the Law Amendment Society and a group lobbying for a woman's right to retain her own property after she married (rather than control passing to her husband). The aforementioned Property Act partly satisfied the latter group. Over the years SSA conferences attracted a wide range of other single-issue pressure groups, creating an impression that it was chasing the latest fad in an undisciplined pursuit of activity for its own sake. As Spencer had noted in *Man Versus the State*, 'the blank form of a question daily asked is – "We have already done this; why should we not do that?"'.[17] For all his interest in social science, Spencer would have nothing to do with the SSA.

---

[17]Offer (ed.), *Political Writings*, p. 88.

Instead of being a melting pot in which principles of social science fermented and crystallized, the SSA seemed to collect divisive special interest groups into a salad bowl. Indeed, it could even be seen to be breeding them. In 1868 the SSA set up a 'Labour and Capital Committee' to discuss trade unions and industrial unrest: members included Ruskin, Gladstone and Ludlow. This effort to knock heads together failed, however, when the workers in attendance noted that they were not being treated as equals. Disgruntled, they left – and formed the Trades Union Congress (TUC).

The demise of the SSA in 1886 can be related to the demise of the Liberal Party with which it was associated. The latter was caused by Gladstone's attempt to pass a bill giving Ireland 'Home Rule', that is, a separate parliament with control over all policy areas except defence and foreign affairs. This attempt failed, and was followed by a general election victory for the Conservatives, led by the third Marquess of Salisbury. Home Rule split the Liberal Party into two. Joseph Chamberlain formed a new, Liberal Unionist Party. This served many former Liberals, including Huxley, as a bridge to the Conservative Party. The Liberal Unionist and Conservative parties ruled in coalition in 1895; a complete merger took place in 1912. Meanwhile the rump of the Liberals limped on as a quarrelsome association, one which almost tore itself apart during the Boer War (an armed rising by Dutch settlers in South Africa against British imperial rule, 1899–1902). Apart from a landslide in 1906, the Liberals never led a majority government after 1886.

In terms of political thought this period saw the emergence of New Liberalism, associated with T. H. Green and L. T. Hobhouse. This was built on the kind of thinking found in Huxley's 'Administrative Nihilism': a sense that the state had a duty to do more than simply police the 'law of equal freedom'. The state could promote moral behaviour. The 'old' Liberals had derived their identity and purpose from the same campaigns that had exercised the Spencers: Disestablishment of the Church of England, free trade, an end to 'Old Corruption' in government and improved access to information (abolition of the 'Taxes on Knowledge', free museum access). Once the working man had a level playing field, 'Old' Liberals reasoned, government could sit back and let 'self-help' take over. 'New Liberals' argued that it was unfair to expect the masses to compete in the 'struggle for life' when they lacked education, sanitation and economic opportunities enjoyed by their superiors.

A new awareness of the economic cycle made it harder to equate unemployment with laziness. There would be times when work was unavailable, or inaccessible, due to the need for training, or illness. Unemployment insurance was necessary to help tide workers over such times. Such insurance was proposed by the Rev William Blackley in 1878 and was considered by a parliamentary select committee in 1885. Although it was only introduced in 1908, for Spencer the message was clear: New Liberalism was a fancy name for Socialism, if not Communism. The weaker, lazier and more irresponsible an individual was, the more state assistance he would receive. Instead of its weaker members being 'excreted' (as Spencer had put it in *Social Statics*), that is, instead of them dying off they would survive and multiply, outbreeding the provident, those who only started families when they knew they could support them.[18] The species would degenerate and eventually be wiped off the face of the earth by a rival species.

Unless, that is, the experts could use the state's expanded powers over its citizens to intervene directly in human reproduction. First mooted by Charles Darwin's cousin, Francis Galton, in his 1865 study of *Hereditary Character*, 'eugenics' (Galton coined the word) made the state the selector: either by encouraging the 'fit' to breed (by offering financial incentives) in 'positive eugenics' or, if necessary, by preventing the 'unfit' from reproducing, in 'negative eugenics'. As we have seen, in the 1830s Malthusian fears of the 'multiplication of the unfit' encouraged the construction of workhouses in which wives and husbands were separated. Darwin had noted the dysgenic effects of human charity towards the weak in *The Descent of Man*, but cut his discussion short, insisting that any other approach would rob mankind of 'sympathy'. Around the turn of the twentieth century, however, declining population, fear of immigrants and a concern at the physical weakness of working-class recruits into the armed forces led to eugenics being taken far more seriously.

The rise of Mendelian heredity noted in the introduction also played a role, by suggesting that harmful traits (morphological ones, but also proclivities such as alcoholism, criminality, feeble-mindedness and so on) could be passed down the generations 'invisibly', that is, through individuals who did not themselves exhibit those traits. The resistance of such hereditary 'taints' to eradication made eugenic vigilance seem particularly urgent. But was this vigilance a state responsibility, or could individuals be left to patrol their own heredity? Darwin's son exemplifies one way Victorians sought to fudge such dilemmas.

UNCLE LENNY AND BRITISH EUGENICS

As a child Lenny's habit of jumping up and down on the sofa caused his father no end of grief. Leonard entered the armed forces in the belief that he was the least intelligent of the great Victorian's brood. He nonetheless served with distinction as an astronomer, chemist and intelligence officer before resigning in 1890. After a brief career as an MP he succeeded Galton as president of the British Eugenics Education Society in 1911, publishing *What is Eugenics?* in 1928. In it he echoed Spencer's concern at the dysgenic effect of state welfare. Rather than adopting Spencer's *laissez-faire*, Leonard argued that the state should incentivize

[18]Ibid., p. 127.

'good stock' to breed by family allowances, 'good stock' being 'all healthy men drawing good wages for useful work done'.[19] Adopting a Mendelian, 'hard' model of inheritance, Leonard saw 'bad' stock as epileptics, the handicapped, diseased, 'feeble-minded,' 'habitual criminals' and those dependent on state welfare – as well as anyone with the 'seeds' of such genetic taints in their family tree.

Although he accepted compulsory sterilization as a last resort, a respect for 'the rights of the individual' led Leonard to prefer education and positive reinforcement. He also saw a role for civil society (including churches) in fostering a sense of patriotic duty, which would encourage healthy men to breed (rather than trying to move up in society), healthy women to marry young (rather than pursue higher education and careers) and the unhealthy to refrain from marriage and reproduction altogether. Eugenics for Leonard Darwin thus incorporated an ethics of self-sacrifice, although Spencer might not be the only one to wonder if the British state merited such a sacrifice, or if Leonard's heady vision of 'rebellion and chaos within and invasion from without' wasn't a bit overblown.

Spencer certainly wasn't the only one to question the Victorian assumption that hard work and money represented 'fitness'. Leonard's niece, the artist Gwen Raverat, recalled debating just this point with 'Uncle Lenny' in her 1952 memoir, *Period Piece*:

> Uncle Lenny used to shock me when, in talking about Eugenics, he maintained that a money standard was the only possible criterion in deciding which human stocks should be encouraged to breed . . . "A man who can earn and keep money shows that he has the qualities essential to survival." I said that money had little importance for such people as artists, philosophers, inventors, gypsies. It would be an irreparable loss to the human race if those valuable strains were to be bred out altogether. It depressed me deeply to think of everybody living in tidy suburban villas, earning and keeping money – what for? But Uncle Lenny could not see this at all. He had very little use for artists; and gypsies were generally dirty and dishonest.[20]

Raverat's memoirs are an invaluable window on the twilight of the Victorians. Raverat shares the family's pride in its heritage (a heritage which Galton – himself a member of the clan – dressed up as the science of heredity) but has a child's detached view of it. She can see how her uncle's eugenics exists in tension with other, unexamined concerns surrounding Intemperance, 'propriety' (contraception is addressed gingerly in *What is Eugenics?*), a bourgeois desire to keep up appearances and 'get on' in the world, and, most importantly, a vestigial 'old' Liberal's belief in individual self-determination.

---

[19]Leonard Darwin, *What Is Eugenics?* (New York: International Congress of Eugenics, 1932), p. 75.

[20]Gwen Raverat, *Period Piece: A Cambridge Childhood* (London: Faber & Faber, 1987), pp. 199–200.

It is impossible to read Leonard Darwin's *What is Eugenics?* today and not be aware that within 5 years the German Parliament would pass a Sterilisation Law, heralding the horrors of the eugenic 'final solution', the systematic rounding-up, internment and murder of Jews in Germany and other states that Nazi Germany controlled. The Nazis certainly agreed with Darwin that Romany (or 'gipsies') were 'dirty' and endeavoured to eliminate them as well. The Sterilisation Law also addressed Darwin's other horrors, targeting those guilty of 'anti-social behaviour' as well as victims of hereditary illness (in which they included alcoholism).

Why was there no eugenic legislation in Britain? One response might be to insist that there *was* in fact such legislation. One could point to the influence of eugenic (and anti-semitic) thought on the 1905 Aliens Act, for example, which introduced immigration controls for the first time. One could also point to the 1913 Mental Deficiency Act, which sought compulsory internment of the 'feeble-minded' and 'moral defectives'. The British Eugenics Education Society campaigned hard for the Bill. Although it passed, it did so with difficulty and with a number of amendments softening its terms. These amendments were the work of a single Liberal MP, Josiah Wedgwood, who sustained himself throughout hours of a House of Commons debate on a stash of chocolate bars. A descendant of the eighteenth-century potter Josiah Wedgwood, he was distantly related to Leonard Darwin and would later serve in the country's first Labour cabinet. Josiah stood for an 'old' Liberal tradition which refused to be seduced either by the power of the state or by science.

This suspicion of the state and fear of 'despotism' (however 'scientific') were the main reasons that Britain never saw the likes of the 1933 Sterilisation Act. Social Darwinism is often blamed for leading to eugenics and Nazi genocide. As a 'Social Darwinist', therefore, Spencer is presumably to be blamed both for libertarian opposition to any state intervention *and* for a despicable abuse of state intervention. Poor Spencer, it seems, just cannot win: he is damned for what the state didn't do and damned for what it did. That 'Social Darwinism' can be associated with both may indicate that the term has become too flexible to retain any tension. In the same way that Spencer's highly deductive approach led to charges that his Synthetic Philosophy offered little more than a redescription of the world, so accounts which defend a 'social darwinist mindset' may simply be defending another term for 'Victorian'.

What do we see if we put this term to one side? Spencer's evolution was heavily influenced by Lamarck, whom Spencer had first encountered, thanks to Lyell's *Principles of Geology*. It reflected Lamarck's belief in use–disuse and the inheritance of acquired characteristics. But it was also uniformitarian in a Lyellian sense, describing slow, incremental processes that were still going on, right now. Like Adam Smith, Spencer was thinking as a moral philosopher. He defined 'human' as a set of instincts that brought individuals together and led them to interact and cooperate for their mutual betterment. Deliberate attempts to instil social passions were doomed to fail. These instincts and the upwards progress of civilization were implanted in humanity and foreordained by God, not that Smith or Spencer wrote about Him much. Combe's phrenology encouraged Spencer's optimistic view of the future 'perfection' of humanity. Admittedly Spencer's views became less optimistic after a nervous breakdown in 1856. The perils of 'degeneration' were portrayed in more vivid colours in *Man Versus the State* (1884) than they had been in *The Proper Sphere of Government* (1842). But Spencer never lost faith in progress.

To be a man of science was, Spencer wrote, to feel 'more vividly than any others can feel, the utter incomprehensibleness of the simplest fact, considered in itself. He alone truly *sees* that absolute knowledge is impossible'.[21] He seems to be confronting his 'facts' in the same way that Huxley does his 'cards' in the analogy discussed in the introduction. Thanks to their shared admiration of German embryology, both Huxley and Spencer emphasized 'wet' and 'soft' morphology over 'dry' and 'hard' osteology. Slippery or even intangible properties like 'mind', 'character' and instinct determined the shape of bones and beaks, not the other way around. Its breathtaking confidence and scope contrasted sharply with Darwin's hesitation and second-guessing of his reader's response.

For many Victorians Spencer's progressive evolution provided reassurance and fuelled aspiration. Huxley's frustration with his close friend in later life was born of a sense that Spencer had let Christian believers off the hook. Spencer allowed them to dress up his 'Unknowable', as it were, with the values and responsibilities they had dressed their Protestant God up in. The lectures, sermons and books of the Oxford theologian Aubrey Lackington Moore (1848–90) and the Scottish Free Church minister James Inverach (1839–1922) certainly used Spencer to advance Kingsley's project of a 'natural theology of the future' into the late 1880s and 1890s. Other Victorians yoked Huxley's own agnosticism to this Providential, 'Darwinian' (actually Lamarckian) theodicy, transforming something Huxley had understood as a scientific tool into a creed or cult. Even more frustratingly for Huxley, advocates of this 'Philosophical Agnosticism' included some of the most effective popularizers of evolution in the High Victorian age, notably the stable of writers around the publisher Charles Albert Watts (1858–1946), editor of *The Agnostic*. These included Richard Bithell, author of *The Worship of the Unknowable* (1889) and Samuel Laing, who wrote a creed for agnostics. These agnostics venerated evolution as a manifestation of His (the Unknowable's) power. Their theodicy was explicitly opposed to socialism, in that it saw evolution as teaching 'individual initiative and enterprise in material life'.[22]

---

[21]Spencer, 'Progress: Its Law and Cause', in Spencer, *Essays*, 1: 62.

[22]Bernard Lightman, 'Ideology, Evolution and late-Victorian Agnostic Popularizers,' in Lightman (ed.), *Evolutionary Naturalism in Victorian Britain* (Farnham: Ashgate, 2009), ch. 7, p. 296.

If Spencer's twentieth-century admirers hailed him as a free-market capitalist, contemporary admirers made the philosopher into something of a guru. These Victorian admirers exhibited what James Moore has called a 'crypto-theistic agnosticism which could become all things to all men'.[23] This loose yet vaguely reassuring belief system arguably persists to this day among the many self-styled agnostics who are 'spiritual, but not religious:' who refuse to accept that the universe is a product of chance and random variation, who hold with some form of Providence, but who recognize no church and no doctrine. Spencer's evolution did not represent a secularist assault on Christianity or belief, therefore. On the contrary, it became a faith in itself.[24]

Further reading

Collini, Stefan, *Public Moralists: Political Thought and Intellectual Life in Britain, 1850–1930* (Oxford: Clarendon, 1991).

Mark Francis, 'H. S. Maine: Victorian Evolution and Political Theory,' *History of European Ideas* 19 (1994): 753–60.

Francis, Mark, *Herbert Spencer and the Invention of Modern Life* (Stocksfield: Acumen, 2007).

Kevles, Daniel J., *In the Name of Eugenics: Genetics and the Uses of Human Heredity* (Cambridge, MA: Harvard University Press, 1995).

Goldman, Lawrence, *Science, Reform and Politics in Victorian Britain: The Social Science Assocation, 1857–1886* (Cambridge: Cambridge University Press, 2002).

Hawkins, Mike, *Social Darwinism in European and American Thought, 1860–1945* (Cambridge: Cambridge University Press, 1997).

Spencer, Herbert, *Political Writings*, ed. John Offer (Cambridge: Cambridge University Press, 1994).

Taylor, Michael, *The Philosophy of Herbert Spencer* (London: Continuum, 2007).

Wiltshire, David, *The Social and Political Thought of Herbert Spencer* (Oxford: Oxford University Press, 1978).

Woodhouse, Jayne, 'Eugenics and the Feeble-Minded: The Parliamentary Debates of 1912–4,' *History of Education* 11 (1982):127–37.

[23]James R. Moore, *The Post-Darwinian Controversies* (Cambridge: Cambridge University Press, 1979), p. 300.

[24]Robert M. Young, *Darwin's Metaphor: Nature's Place in Victorian Culture* (Cambridge: Cambridge University Press, 1985), p. 240.

CHAPTER EIGHT

Domestic evolution? Making a home for science

On 29 November 1859, just a few days after the publication of *The Origin*, Darwin wrote to Huxley from Ilkley Wells, where he was once again undergoing a series of water treatments. As we have seen, Huxley had read and found the book's hypothesis a highly stimulating one. He asked Darwin to recommend books to read on artificial selection, recognizing the importance Darwin had placed on evidence from breeders in his book. Darwin was eager to help, but noted the difficulty in finding reliable printed sources. Publications issued by county agricultural societies could provide conflicting information, he noted. Happily, non-printed sources, living experts could be consulted.

But only if one was willing to slum it:

> I have found it very important associating with fanciers & breeders. – For instance I sat one evening in a gin-palace in the Borough amongst a set of Pigeon-fanciers, – when it was hinted that Mr Bull had crossed his Powters with Runts [two types of pigeon] to gain size; & if you had seen the solemn, the mysterious & awful shakes of the head which all the fanciers gave at this scandalous proceeding, you would have recognised how little crossing has had to do with improving breeds, & how dangerous for endless generations the process was.[1]

---

[1]Darwin to Huxley, 27 November 1859. Darwin Correspondence Database, http://www.darwinproject.ac.uk 2558 (accessed 2 May 2013).

Darwin's expedition to a gin palace was daring in more ways than one.

By the 1850s pubs had become off-limits to gentlemen. Despite their grand name and lavish interiors (full of mirrors and brass), gin palaces were no more respectable. The Victorian campaign against alcoholic drink, the Temperance movement, encouraged men and women of Darwin's class to view gin palaces as dens of corruption. In George Cruikshank's illustrated fable, *The Drunkard's Children* (1848), we follow the downward spiral of a young boy and girl, who lose their parents to alcoholism only to be tempted and fall themselves. The scene showing them in a gin palace includes vignettes intended to shock readers into 'taking the pledge', that is, signing a promise to abstain from alcohol entirely. A ragged mother gives her baby gin to keep her quiet. By the bar a man desperate for his next fix impersonates a dog in order to earn a few pence from fellow lowlifes. The heroine is being chatted up by a pimp. In the next scene we see her, ruined, jumping to her death from a bridge. What on earth was a learnt gentleman like Darwin doing in such an establishment – in Borough, a disreputable part of south London most middle-class Londoners only knew from reading Charles Dickens?

Darwin was going to learn. The knowledge he gained was not written down. It was peer-reviewed, but the peers involved were not fellows of a 'learned society', at least, not in the sense that Darwin and Huxley were Fellows of the Linnean and Geological Societies, London-focused societies with proud lineages, high annual subscription fees and membership drawn from a highly educated, national elite. It was passed down, discussed and shared orally; there were no published *Transactions* or *Proceedings* such as those issued by the Royal and Linnean Societies. It was not very articulate. Indeed, Darwin notes that it was the dumb show, 'the solemn, the mysterious and awful shakes of the head', which most impressed him. This response convinced him that Huxley and others were wrong to assume that the many varieties of pigeon proudly displayed at pigeon shows were the result of crossing between varieties, rather than being the result of a much longer, incremental process of artificial selection for this or that trait among the chicks born to pairs of runts (or pairs of pouters, jacobins and so on). This carried more authority for Darwin, he continued, than 'eight pages of mere statement and etc'.

Gin palaces and pubs played host not only to pigeon-fanciers, but also to other groups of working men with an interest in botany, geology and other natural sciences, in Lancashire mill towns as well as in the capital. These gatherings were confusing to those writers eager to celebrate natural history as a *means* (of encouraging 'moral improvement') rather than as something directed towards an *end* (the advancement of knowledge). For middle-class 'improvers', natural history was 'rational recreation', a means by which the humble working man could be kept out of trouble while learning to admire the wonders of Creation. The fact that such gatherings usually occurred on Sundays was even more troubling, given the Sabbatarian movement to 'keep the Sabbath day holy', by attending church (twice) and staying quietly at home. Members of middle-class Sabbatarian groups organized prosecutions of publicans and theatre managers who operated on Sundays. In November 1850 Joshua Barge, keeper of The Ostrich in Prestwich, was fined £5 for serving drinks to a group of botanists.

The Ostrich had been hosting such meetings and sponsoring prize gooseberry shows for years, part of a tradition of working-class botany clubs stretching back into the previous century.[2] Together with Mechanics Institutes like that established in Neath by Alfred Russel Wallace the previous decade, these clubs had small libraries, providing members with the chance to borrow books they could never afford to buy on their own. This was an important service, particularly in the years before the 1850 Libraries Act facilitated the foundation of municipal public libraries, when the only libraries were commercially operated ones such as Mudie's (subscription: 1 guinea a year, equivalent to several week's wages for a clerk). Though they defied middle-class notions of how and where 'useful knowledge' was supposed to be 'diffused', such clubs were not unruly. They had their own system of membership fees and rules (including one fining members who showed up drunk), and reflected a rich working-class culture centred on the public house as a social centre (not just a place to buy and consume alcoholic beverages).

Our knowledge of how the pub operated as a 'home' for science is limited by the lack of records. The autobiographies of working-class men are useful, but can be heavily coloured by accounts of religious conversion and a certain aloof pride in having been marked out from one's peers by a love of reading or the pursuit of a particular scientific interest. Writing in 1866 one radical jeweller named Morrell contrasted the 'coarse and ignorant' habits of the working man in his youth, when Sundays were spent wrestling and staging dog-fights in the street, with 'the purer and more rational enjoyment' later offered by free public museums, Kew Gardens and cheap excursions to the country or seaside.[3] Such accounts did little to challenge middle-class notions of working-class recreation as a threat, as so much energy in need of channelling towards physically and morally healthy 'counter-attractions' like reading, museum visiting and nature walks.

This desire to 'improve' the working man explains middle-class and elite support of Mechanics Institutes and the Society for the Diffusion of Useful Knowledge (1826). In addition to keeping men out of the pub and teaching them 'home-loving habits', it was believed that working men would use the knowledge they gained to make practical improvements to the industries they served, advancing the economic fortunes of the nation as a whole.

---

[2]Anne Secord, 'Science in the Pub: Artisan Botanists in early nineteenth-century Lancashire,' *History of Science* 32 (1994): 269–315 (275–6).

[3]Cited in Iorwerth Prothero, *Radical Artisans in England and France, 1830–70* (Cambridge: Cambridge University Press, 1997), p. 282.

This was a Whig ideal of the 1830s, one which refused to share earlier Tory fears that such activities were a cover for 'French principles', illegal 'combinations' (i.e., trade unions) and political insurrection. By the 1850s, such hopes had largely evaporated. The Society folded in 1848. Mechanics Institutes shifted their focus to entertainment. Fearful of working-class participation in scientific activity and their appropriation of knowledge, the middle class made 'popular science' into a neutral field for 'harmless and industrious' activity, only worth celebrating in terms of the virtues of self-help and discipline that it instilled in its practitioners, not in terms of their contribution to knowledge.

The working men who advised Darwin had to teach themselves because there was no system of state education. They would have received basic training in literacy and numeracy in village schools (where a small fee was paid by parents) or in the free 'Ragged Schools' which began appearing in London and in other large towns in the 1840s. Schooling would, many believed, only delay boys' and girls' entry into the workforce, which normally happened around the age of 11, making them a burden on the state and their families. Thus the first halting steps towards state education in the 1830s were an offshoot of state regulation of factories, of health and safety legislation (to use today's term).

Under the 1833 Factory Act children under the age of 9 were prohibited from working, while those child workers aged between 9 and 13 were to have 2 hours' teaching a week. Meanwhile any newspaper publishing articles on current affairs was required to print on specially taxed paper, a much-resented 'Tax on Knowledge', which also limited working-class access to politics. This tax was repealed in 1855, while in 1870 came Forster's Education Act. The Act set off a wave of school-building activity and the establishment of the first Boards of Education, which received their more familiar name of Local Education Authorities in 1902. In 1882 school attendance was made mandatory for children between the ages of 5 and 12.

Although Darwin's trip to the gin palace helped him develop his ideas, it was unusual for men of his class to venture into such places. He clearly expected Huxley to find the idea of his friend visiting such a dive amusing. Botany, pigeon and other working-class societies had their own set of conventions, but their knowledge was not widely shared. Only Darwin, it seems, saw the connection between their activities and the debate over transmutation. Though he drew on pigeon-fanciers' knowledge himself and encouraged Huxley to illustrate his lecture with a display of pigeons, Darwin treated the fanciers differently than he did the many middle- and upper-class men whom he constantly badgered for nuggets of information. He advised Huxley to pay them for the loan of their pigeons, either in money or in tickets to his lecture. This clearly established the relationship as a business transaction between a client and a tradesman. 'Gentlemen of science' assisted one another without money changing hands. Indeed, had Darwin offered to pay his genteel correspondents for the information they provided, the latter would have been highly offended.

There were other spaces in which knowledge of natural history could be gathered, exchanged and discussed in Victorian Britain. These spaces were more 'public' than Darwin's pub, in that they were accessible to different classes, to men and women, young and old, to local residents as well as to those further afield. They were also more public in their pretensions to serve the common interest, by promoting a health-giving interest in 'useful' or 'practical' sciences, by instilling virtuous habits and by inspiring patriotic pride in *British* science. A few were state-funded, by municipal or central taxation. Many of these institutions were middle-class initiatives aimed at providing the working man with 'rational recreation', that is, with alternatives to drinking, gambling, bull-baiting and other pursuits.

Though the rest of this chapter will see us exploring more familiar scientific institutions such as lecture halls, menageries and museums, Darwin's gin palace serves as a salutary starting place, as it reminds us that the working classes were perfectly capable of forming their own scientific societies without the help of their betters. In considering these spaces for Victorian science we not only need to consider the rules, hierarchies and behavioural conventions observed within them, but also need to consider the allusions and cues given by their architecture and layout. Alongside the built environment we will be considering the printed page, a virtual space of discovery and encounter.

To what extent was the natural history encountered within these spaces an evolutionary one? Broadly speaking Victorian popular science was about exercise, humility and reverence. The study of geology was about getting out of the house and tramping across fields and along cliffs, as well as mental exercise. Astronomy instilled humility by contrasting one's diminutive self with the vastness of the cosmos. Body-focused appetites and desires were subdued as observation and study literally led one 'out of oneself'. Finally, all natural sciences encouraged reverence for a divine creator by noting the presence of order and purpose amidst the extent and diversity of the universe.

By this account one might be excused for seeing popular science as opposed to evolution. As we shall see, Victorian natural history museums largely reflected the natural theology of Paley's era. Otherwise, however, these institutions did not so much oppose as tame evolution. Significant concessions were made. As the century progressed the entity towards which this reverence was directed became ever more diffuse, becoming a collection of attributes without a 'person' to whom they could be attributed. The mental and physical effort expended on natural history was linked to the 'struggle for life'. Disturbing and even frightening encounters with the evolutionary past were certainly not ruled out. Indeed, the London Zoo and Sydenham dinosaurs used the wonders of modern transportation and construction to bring them to life. Queen Victoria was not the only visitor to be disturbed by the obvious similarities between her species and the residents of the Zoo's monkey house. At the Crystal Palace in Sydenham England's fearsome extinct megafauna came back to life and roamed south London.

Reading and rambling

Improvements in printing technology, enterprising publishers and a large literate public (by European standards) meant that Victorians enjoyed a wide range of published work on natural history. The same publisher, even the same author could produce works intended to serve very different audiences and purposes. Reading such texts in print-on-demand versions or online, it is very easy to overlook or miss the clues which indicate the kind of audience a particular text is intended for. In addition to the title and the words on the page we also need to consider register and tone, considering whether technical or colloquial language is being used, for example, whether the reader is being lectured at or invited to take part in a friendly conversation. We also need to consider price, information that can be hard to get at. Illustrations often provide a quick way of gauging price, but need to be assessed in the light of rapid improvements in engraving, lithography, chromolithography, talbotypes (an early form of photograph) and other techniques of reproduction.

Happily, original copies of bestselling works from the 1870s onwards are relatively easy to find in second-hand stores, usually for a few pounds. These often contain advertisements for other works of natural history, often with information on 'deluxe' or 'collectors' editions, that is, editions which might include gilded rather than plain paper edges, decorative bindings and colour (rather than cheaper black and white) illustrations. Some contain dedications or bookplates indicating that they were gifts or Sunday School prizes, or formed part of the stock of a college's or Working Men's Institute's library. Such books may have 'uncut pages', that is, the front edges of some pages may need opening with a penknife. Taken together, this evidence can indicate who bought the book and why, and whether they in fact read it. The study of these aspects, of the 'history of the book', has recently come to interest scholars, even as the book-as-object becomes less familiar.

Printed fieldguides were particularly profitable for publishers, if not necessarily for their authors. Charles Alexander Johns' *Botanical Rambles* (1846) set a trend for authors to organize their works around a series of virtual jaunts in the company of the author, who writes in the first person. Kingsley followed this model in *Glaucus, or the Wonders of the Shore* (1855) and *Madam How and Lady Why* (1870) as well as his *Town Geology* (1877). Rhetorical questions and answers lend these works a chummy feel which today's readers can find grating. Where we may feel ourselves being manhandled across fields and along beaches like a ventriloquist's dummy, Victorian readers were charmed and reassured that this natural history thing was in fact within their reach. Reassurance was necessary precisely because even those who were very well educated by Victorian standards would not have received any education in the natural sciences. They did not know what to expect.

Authors of such guides steered well clear of controversy, following the example set by the first introduction to seaside natural history, William Harvey's *Book of the Sea-Side* (1849). Echoing the spirit of the Bridgewater Treatises, Harvey warned the reader against speculation and system building. Harvey, Philip Henry Gosse, Johns and their rivals were endeavouring to make natural history safe, and 'safe science' did not try to prove the existence of God. Though they did regularly refer to this or that feature or trait as a *product* of divine wisdom and ingenuity, it was a casual, sometimes formulaic reverence, totally lacking in that stridency and insistence one might have expected, had these features been presented as *evidence* of divine wisdom and ingenuity. As Lightman has noted, such writing embodied a 'theology of nature', not a contribution to 'the demonstrative natural theology as presented by Paley'.[4]

When Johns and Yarrell thus wrote of how, say, the beak of the crossbill (a bird) was 'exquisitely adapted to its work', it went without saying that the credit was due to God's ingenuity. Of course, a strict Darwinian could read the same passage and find nothing objectionable, even if she saw natural selection as doing the adapting. Was this commercially minded prevarication a case of authors deliberately using equivocal language in order to attract the widest possible audience? Given the consensus view, which admitted a good deal of transmutation without worrying too much about its broader implications, this seems unlikely. Language like this is very common in such books and may not be as tactical as it may sound to us today.

What these authors did for the 'wonders' of the earth, the Rev Thomas Webb did for those of the heavens, as the father of amateur astronomy. Webb's *Celestial Objects for Common Telescopes* (1859) echoed their reverence for the wisdom of God. So keen was Webb to emphasize the size and complexity of the observable universe that he made astronomy seem almost pointless. If, as Webb noted, the most experienced observers could not agree on the features of Mars, if the most sophisticated spectroscope offered inconclusive evidence on the substance of nebulae, why bother? Webb's astronomer studied the heavens with his telescope as a worthy activity, one pursued for its own sake, or rather for the values and habits it instilled. On this basis disagreement among earlier writers or men of science over, say, transmutation or the nebular hypothesis could be dismissed as a petty squabble of little real interest. This, too, reassured the general reader, encouraging him to dip his or her toe in natural history.

---

[4]Bernard Lightman, *Victorian Popularizers of Science. Designing Nature for New Audiences* (Chicago: University of Chicago Press, 2007), p. 24.

So did satirical depictions of set-piece debates like that between Huxley and Owen over the 'Hippocampus Question', satires such as those published by Kingsley and an anonymous writer in 1862 and 1863, respectively. It was hard to take a debate over the common ancestry of ape and human entirely seriously when it was written as a farcical judge-and-jury show, with Huxley depicted as a cockney stallkeeper struggling with an older rival ('Dick,' i.e. Richard Owen) for control of the market in 'old bones, bird skins, offal, and what not':

> Well, as I was saying, Owen and me is in the same trade; and we both cuts up monkeys, and I finds something in the brains of 'em. Hallo! says I, here's a hippocampus. No, there ain't, says Owen. Look here, says I. I can't see it, says he, and he sets to werriting and haggling about it, and goes and tells everybody, as what I finds ain't there, and what he finds is, and that's what no tradesman will stand. So when we meets, we has words.[5]

The satire worked because the dispute could indeed be seen as the result of Owen's desire to be top dog. As Darwin wrote to Huxley in 1859, 'credit given to any other man, I strongly suspect is in [Owen's] eyes so much credit robbed from him. Science is so narrow a field it is clear there ought to be only one cock of the walk!'[6]

In contrast with the gentle satire of such Hippocampus parodies, the Rev Frances Orpen Morris' shilling pamphlet *All the Articles of the Darwin Faith* (1875) is mean and cynical. Dedicated, apparently with permission, 'to the Right Honourable The Common Sense of The People of England', it was constructed along the lines of the Church of England's Thirty-Nine Articles, which list Anglicanism's defining beliefs. Morris selectively quoted from Darwin, turning Darwin's characteristic humility against him. As he recites 'his' creed, therefore, Darwin is made to utter some silly things:

> I believe that the sudden appearance of whole groups of species in some strata by no means overturns my doctrine of slow descent by natural selection, though Sedgwick, Agassiz, and others have maintained that it does. What does anyone but myself know about the matter? Geology as at present understood must be false if my Theory is true, and as my theory must be true because it is mine, e'en so let it be. *Fiat Evolutio, ruat cœlum.* [Latin: 'Let evolution be, or let the sky fall'][7]

---

[5][Kingsley], *Speech of Lord Dundreary in Section D. On Friday Last. On the Great Hippocampus Question* (London: Macmillan, 1862); Anon., *A Report of a Sad Case, Recently Tried before the Lord Mayor, Owen vs. Huxley* (London: n.p., 1863).

[6]Darwin to Huxley, 28 December 1859. Darwin Correspondence Database, http://www.darwinproject.ac.uk 2611 (accessed 2 May 2013).

[7]F. O. Morris, *All the Articles of the Darwin Faith* (London: William Poole, 1877), p. 13.

Despite its crustiness *All the Articles* managed to rack up three editions in 7 years. But this needs to be compared to the far greater success of Morris' non-polemical handbooks. Even his multi-volume *History of British Moths* (1859–70) outsold such squibs. Morris knew what the market wanted and wisely kept the vitriol out of his books. Though he shared Morris' belief that Darwin's theory was based on faulty logic, William Harvey regretted publishing a similar 'serio-comic squib' in 1860 and tried to suppress it. Not only was Morris foolish to suggest that Darwin was slavishly followed by 'acolytes', his prolific pamphleteering and shaggy-dog stories (published as *Dogs and Their Doings*, 1872) made him look like a bit of a crank on the subject.

The fad for seaside natural history in the early 1850s started by Gosse encouraged publishers to give the *Rambles in . . .* and *Common Objects of . . .* treatment to other areas of the natural world, as well as to re-publish popular works in new editions, with more illustrations. Wood's *Common Objects of the Sea Shore* and *Common Objects of the Country* (1857 and 1858 respectively) sold 64,000 copies within a decade. While Wood went on to write a grand total of 24 books of natural history, it was his publisher, George Routledge, who profited, leaving Wood struggling to make a living on the British and American lecture circuit, a gruelling routine whose rigours contributed to his death in 1889. In the 1870s and 1880s Longman, John Murray, Macmillan and other publishers launched similar 'series' or 'libraries'. Among them were Christian organizations established in the seventeenth and eighteenth centuries to evangelize: the Society for Promoting Christian Knowledge (SPCK) and the Religious Tract Society (RTS). The RTS had taken its first steps in natural history in the mid-1840s. The SPCK launched its 'Manuals of Elementary Science' and 'Natural History Rambles' series in 1873 and 1879, respectively. New technologies such as hot-metal typesetting and rotary printing made it quicker, easier and less expensive to publish these works.

Many of these works appealed to younger readers. Whether they did so in isolation, however, can be a difficult question to answer. Wood's *Boys' Own Book of Natural History* (1861) was clearly aimed at children, but in other cases neat categorization of a given work as 'children's literature' is difficult. Just because a work has the word *Fairy* in the title, say, or is dedicated to the author's child does not mean that it was exclusively for children, as Kingsley's *Water-Babies* demonstrates, and could not be appreciated by adults. There were several female authors who seem to have specialized in writing for a young audience, including Arabella Buckley (1840–1929), whom we met in the previous chapter. Buckley had been secretary to Lyell and turned to writing after his death in 1875. Her *Fairy-Land of Science* (1879) went through many editions, using a didactic fairy-tale format familiar from Kingsley's work as well as from the bestselling series *Parables from Nature* (1855–71), written by Margaret Gatty (1809–73). Buckley shared Gatty and Kingsley's view that the natural world taught morality. Unlike Gatty, however, who was convinced that *The Origin* would soon be unmasked as 'a great man's *blunder*', Buckley was an evolutionist. She continued to expound a Kingsleyan evolutionary ethics on into the 1890s.[8]

Octavo-sized, with gilt-edged pages and 100 black and white illustrations Buckley's *Life and Her Children: Glimpses of Animal Life from the Amœba to the Insects* was published by Edward Standford in 1880, and cost 6 shillings. My copy is inscribed 'From Tom to Sis' and bears all the signs of heavy use. In the preface Buckley thanks R. Garnett of the British Museum (BM), a Mr Lowne (a Fellow of the Royal College of Surgeons) and a Mr Haddon, demonstrator of Comparative Anatomy at Cambridge University – all men, in posts or with distinctions (such as the FRCS) closed to women. In doing so Buckley does not merely express her gratitude, as all authors do in such prefaces, but also reassures her readers. The unidentified 'Tom' probably found it reassuring to know that the book he was giving his 'sis' was vetted by these gentlemen of science. The quotations from familiar English poets also reassured. An author who quoted Coleridge's lines describing how 'the dear God who loveth us' made 'All things great and small' could hardly be a dangerous materialist. Indeed, Coleridge's lines inspired the Church of England hymn 'All Things Bright and Beautiful', a joyful celebration of the Creator's loving bounty. Yet Buckley herself never refers to God or the Creator in the main text of her book, instead referring to female embodiments of 'Nature' or 'life'.

Her aim, Buckley writes, 'is to acquaint young people with the structure and habits of the lower forms of life; and to do this in a more systematic way than is usual in ordinary works on Natural History, and more simply than in text-books on Zoology'.[9] Buckley seems to be aiming at a middle ground between the 'ramble'-style text (what she calls 'Natural History') and more 'systematic' works aimed at the serious student (instead of 'Natural History', we have the more impressive 'Zoology'). Interestingly, she seems to see all three genres as accessible to 'young people', further emphasizing the extent to which such writing had yet to focus on the latter in isolation. Buckley does not address her reader directly, as Kingsley and other 'ramble'-writers so often did.

Buckley is candid enough to tackle superfecundity head on, noting how many thousands of creatures are born only to die in the 'struggle for life'. She argues that it would not be better 'if only enough were born to have plenty of room and to live comfortably', as

> it is the struggle for life and the necessity for work which makes people invent, and plan, and improve themselves and things around them. And so

---

[8]Lightman, *Popularizers*, p. 156.

[9]Arabella Buckley, *Life and Her Children: Glimpses of Animal Life from the Amœba to the Insects* (London: Edward Stanford, 1880), p. v.

> it is also with plants and animals. Life has to educate all her children, and she does it by giving the prize of success, health, strength, and enjoyment to those who can best fight the battle of existence, and do their work best in the world . . . . If the ocean and the rivers be full, then some must learn to live on the land, and so we have for example sea-snails and land-snails.[10]

Each animal's existence varies 'according to the kind of tools with which life provides it, and the instinct which a long education has been teaching to its ancestors for ages past'.[11] Buckley's account certainly challenges Paley in her cheery Malthusianism, but otherwise her language is ambiguous. On the one hand 'long education' implies slow change; on the other, 'tools' are given to this or that creature, as if by a creator God.

There was a widespread consensus, one which many women shared, that women lacked the 'genius' to think up new ideas of any kind (poetic, artistic or scientific), but were excellent 'copyists' or 'translators' of men's ideas. This combined with their maternal instincts was supposed to make them particularly talented at explaining complex ideas to children. Jane Marcet had blazed a trail in the works she had published on chemistry, economics and physics in the first half of the century, works scripted as conversations between a teacher and her young female pupils. Such writing deferred to male authorities. Gatty described her *History of British Seaweeds* as a primer for readers not yet able to tackle Harvey's *Phycologia Britannica* (1846–51). If these women were trying to establish themselves professionally as popular science writers, they did not succeed. Few made a living of it. Some did it to fill the time, others because they were married to men of science. Although the education of women as doctors was a live issue in the closing decades of the century, otherwise there was little connection between writing on science and campaigning for women's suffrage.

From menagerie to music hall

For centuries the crowned heads of Europe had collected exotic beasts, giving and receiving them as diplomatic gifts and displaying them in the grounds of palaces or, in Britain's case, the Tower of London. Little thought, however, was given to the average visitor; the aim was to display the king's power; to awe, not to educate his subjects. There were commercially operated collections of living animals, too, which regularly swopped animals with circuses. Those creatures that survived the British weather and regular changes of makeshift accommodation, such as the Indian elephant Chunnee,

---

[10]Ibid., pp. 5–7.
[11]Ibid., p. 8.

were admired more for the tricks they could perform than as specimens of the natural history of Asia.

Brought to London around 1810, Chunnee was the star of Edward Cross's menagerie near Charing Cross, London, where he entertained the crowd by picking up and returning visitors' sixpences and hats. Unfortunately Chunnee ran amok in 1826 and had to be put down: a bloody exercise which took a firing squad and a harpoon. His skeleton ended up on display in Owen's Hunterian Museum. In 1831 Cross acquired a site south of present-day Elephant and Castle and landscaped it as a pleasure ground, dotted with pavilions in Turkish, Chinese and other exotic styles. Surrey Zoological Gardens afforded his creatures more room to roam, while visitors enjoyed promenade concerts and dining facilities unavailable in the cramped central London site. Patronized mainly in summertime, Surrey Zoological Gardens was a development of the eighteenth-century pleasure garden: like its more famous south London rival, Vauxhall Gardens, it afforded Londoners a pleasant suburban retreat as well as a place to show off, to see and be seen.

But what of zoos? The word 'zoo' derives from a popular abbreviation used to refer to the Zoological Society of London (ZSL) and its menagerie in London's Regents Park. Established in 1826, the ZSL was the brainchild of Sir Stamford Raffles, a British imperialist responsible for establishing Singapore as a hub for British trade and influence in Southeast Asia. As noted in Chapter 2, Paris was then felt to be far ahead of London in its provision for the natural sciences. Raffles was inspired by an 1817 visit to the Jardin des Plantes, which contained the remains of the French royal menagerie, alongside its other natural history collections. In 1824 he raised the idea of establishing a London equivalent with the president of the Royal Society, the chemist Humphry Davy. A body of influential aristocrats and learnt men was duly formed, with rooms in fashionable Hanover Square in which to display prepared specimens and hold meetings. Research papers were read and discussed, then published in the Society's *Proceedings* and *Transactions*, just as the Linnean and Royal Societies did. The Society's menagerie opened in Regents Park in 1828 and was initially open only to members and their guests.

Since the late twentieth century the institutions established in imitation of ZSL have justified their existence in terms of conservation. It is striking, therefore, to note that among the first zoo's original aims was 'the introduction of new varieties, breeds and races of animals for the purpose of domestication or for stocking our farmyards, woods, pleasure-grounds and wastes'.[12] Far from seeking to protect natural diversity from human intervention it was engineering it to serve the British empire. The Royal

[12]R. Fish and I. Montagu, 'The Zoological Society and the British Overseas,' in Solly Zuckerman (ed.), *The Zoological Society of London* (London: Academic, 1976), pp. 17–48 (28).

Society had led the way in its attempt to organize the transfer of the Tahitian breadfruit to the Caribbean, where it could provide a cheap food for British slaves working in the sugar plantations. Unfortunately the first attempt (1787) ended disastrously when the crew of *HMS Bounty* mutinied. The second attempt succeeded, but the project nonetheless failed when the slaves refused to eat the fruit.

This focus on what might be called 'economic zoology' emphasized the duty of man in Genesis to 'subdue' Creation, reflected a sense that all other forms of life were there to serve man (and here 'man' did indeed mean human males, rather than humanity or mankind). Regions which did not serve man were labelled 'wastes', blank pages which needed filling by man. A child of this Paleyite world, Darwin shared this view; his delight in attempting to eat the creatures he bagged on *HMS Beagle* is hard to explain in any other way. Whether the creature concerned was a Galapagos tortoise or a Patagonian rhea, Darwin had a tendency to eat first and ask questions later. A founding member of ZSL, William Yarrell's two-volume *History of British Fishes* (1835–36) was a pioneering work of ichthyology. To today's ichthyologists, however, Yarrell's investigations into the comestibility (eatableness) of every fish are inexplicable (who cares?). ZSL's early experiments with breeding carp and mammals at a farm in Kingston in order to help its aristocratic members stock the grounds and ponds of their country estates seems equally 'unscientific'. Yet they reflected Raffles' frustration at 'how few amongst the immense variety of animated beings have been hitherto applied to the uses of man'.[13] Founded in 1860, Frank Buckland's Acclimatisation Society pursued a similar aim.[14]

Like the RCS, the ZSL could view its collections in anthropocentric terms, therefore. Just as Owen struggled to justify the Hunterian's holdings of material unrelated to surgery, so the ZSL struggled to justify its 'dead' collections, which were either sold or given to the BM in 1855. Meanwhile its 'living' collections expanded rapidly, especially after the general (paying) public were admitted in 1847. Then as now, it had to find a balance between the systematic collector's ideal of a fully representative collection and a showman's eye for the bottom line. Charismatic megafauna like the hippo (acquired 1850), black rhino (1868) and gorilla (1887) drew crowds, as did lions and monkeys. Insects did not, which is why construction of an Insect House had to wait until 1881. The first purpose-built Monkey House opened much earlier, in 1839, and gave the public direct access: visitors were permitted to feed the monkeys, some of whom were displayed chained to vertical poles. Prodded and poked, they occasionally lashed out, tearing the fringes off ladies' dresses.

---

[13]R. V. Short, 'The Introduction of New Species of Animals,' in Ibid., 321–33.

[14]Bernard Lightman, 'Frank Buckland and the Resilience of Natural Theology,' in Lightman (ed.), *Evolutionary Naturalism in Victorian Britain* (Farnham: Ashgate, 2009), p. 8.

THE SYDENHAM DINOSAURS

Joseph Paxton's 'Crystal Palace' in London's Hyde Park not only served to house the 1851 Great Exhibition of the Works of Industry of All Nations, but it was also a much-admired exhibit in itself, demonstrating the architectural potential of iron and glass to the exhibition's six million visitors. Although it was only intended to be a temporary structure, when the exhibition closed there was such an outcry at its proposed destruction that a permanent home was sought. In 1854 it reopened in south London, on Sydenham Hill, now operated by a Crystal Palace Company, which intended to use it for displays of science and art, concerts and other events.

Considerable capital was sunk into the endeavour, including £13,729 (£800,000) towards the creation of a series of ponds and islands (Fig 5) intended to allow visitors to 'walk back through time', encountering life-size models of the strange beasts which roamed the earth at different points in the earth's past.

Great care was taken to sit the models on the appropriate stratigraphic level, on rock which matched their point in geological time. The Crystal Palace company sought advice from Gideon Mantell on how the dinosaurs and other extinct megafauna should look; but Mantell refused on grounds of health. The door was open for the younger Richard Owen to grab the opportunity to present himself as the father of the 'dinosaur', the term he had coined in 1841 for Mantell's reptiles.

FIGURE 5 *George Baxter,* The Crystal Palace and Gardens, *1854. The display of life-sized models within the dinosaur park at Sydenham in South London brought Victorians face to face with the former denizens of Surrey.*

Like Mantell, Owen opposed transmutation, and he modelled the Sydenham iguanodon as a compact, heavy pachyderm, standing solidly like a rhinoceros. In the guidebook he wrote Owen described it as 'a herbivorous reptile, of colossal dimensions, but belonging to a distinct and higher order or reptiles, more akin to the crocodiles.'[15] Waterhouse Hawkins, who cast the models, even added a horn to its nose. A step too far for Owen, it nonetheless remained. A highly developed lizard, iguanodon reflected Owen's belief that periods of development had been interrupted by periods of degeneration, degeneration which in this case had turned terrible, massive lizards into the smaller ones of latter-day earth. Drawing the bones (held in various collections) to scale on tracing paper and cutting them out, Huxley later assembled it differently in his iguanodon notebook: as a bird-like creature on two legs, halfway between reptile and bird.

Constructing the iguanodon took 600 bricks, 1550 tiles and 38 casks of cement, making it something of a technical or engineering wonder, as well as a massive illustration of Cuvierite palaeontology. The Crystal Palace Company played up both aspects at a well-publicized dinner held on New Years' Eve 1853 – half-inside the massive mould Hawkins had used to cast the iguanodon. This echoed dinners held inside Brunel's Thames Tunnel (1843) and Matthew Cotes Wyatt's massive bronze statue of Wellington (1846), then the largest bronze statue in Europe. Owen was invited to sit at the head of the table, the place of honour. A cartouche with the name 'Owen' hung nearby, alongside one labelled 'Cuvier'.

Unfortunately plans for this area of the park proved overly ambitious, and the mammoth was only half-complete when Queen Victoria opened the resort in 1854. Construction stopped the following year. Huxley would attack Owen's construction of iguanodon in the late 1860s, proposing his own, bird-like, two-legged stance. But the models remained unchanged: a popular success, if not a financial one for the Company, they helped generations of Londoners to imagine the giants that had once walked in Sussex. They did so more effectively, one suspects, than mounted skeletons or even the 'dramatic taxidermy' (i.e. displays of stuffed animals in naturalistic poses, with painted backgrounds) pioneered in the United States from 1860 and 1880 respectively. It would take another century for the 'dinosaur park' model to establish itself elsewhere. Although the Sydenham models were restored in 2002, today this type of scientific theme park is far more popular in Germany and the United States.

The 1870s and 1880s did not just see a shift in the commercialization of print, it also saw the commercialization of leisure. Music halls had originated in the 1840s and 1850s as individual speculations on the part of publicans, who simply built a large room onto the back of their pub. The performers were often semi-amateurs, unable to find enough bookings

---

[15]Richard Owen, *Geology and Inhabitants of the Ancient World* (London: Bradbury and Evans, 1854), p. 17.

to support themselves without holding down another job on the side. With the construction of much larger (1,000-plus capacity) music halls, however, control passed to management companies with shareholders, and to booking agents who maintained a large roster of 'acts', from singers and sketch-artists to mesmerists and wolf-tamers. A national circuit emerged, with the odd foray to France and the United States. Evangelical doubts concerning the morality of actors and especially actresses (all prostitutes, supposedly) had made the stage seem a parasite in the 1830s, a highway to hell. Astonished by the sheer variety of performances and impressed by the vast sums, ingenuity and technology invested in such shows, 50 years on the Victorians accepted that an entertainment industry was here to stay. Bodies like the National Vigilance Association established in 1885 to police 'public immorality' seemed like so many prudes on the prowl.

As we have seen, the ZSL was not unaware of the commercial side of its operations, and the rival Surrey Zoological Gardens was almost exclusively commercial in nature. The Polytechnic Institution established in Regent Street in 1838 and the Royal Panopticon of Science and Art which opened on Leicester Square in 1854 both have impressive-sounding names, and both claimed a mission to improve 'Practical Science'. The institutions had ornate facades decorated with busts of Davy and other heroes of the natural sciences. From the beginning, however, the interactive displays, demonstrations and performances which took place in these impressive, top-lit halls mingled entertainment and 'useful knowledge'. In 1854 the chemist John Henry Pepper (1821–1900) became manager of the Polytechnic Institution, and soon became a popular lecturer on optical illusions, notably 'Pepper's Ghost', as well as pyrotechnics and engineering. Although Pepper was a charismatic speaker, much of his rhetoric was showman's patter, that is, 'voiceover' accompanying displays of dioramas, performing birds and magic tricks. After 2 years the Panopticon added a circus ring, reopening as the Alhambra, a music hall complete with a promenade bar and cancan dancers.

By the 1870s the divide between 'useful knowledge' and entertainment had entirely disappeared, and not just in London. The Brighton Aquarium, a vaguely Gothic Revival pavilion designed by Eugenius Birch, opened in 1872. It had a reading room, restaurant and conservatory as well as tanks full of shrimp, fish and molluscs. A roller skating rink was later built on its flat roof terrace. In the 1880s it was booking a wide range of variety acts we would normally associate with the music hall: burlesque dancers, pantomimists, 'living statuary', ventriloquists, equilibrists and 'Japanese illusionists'.[16] With William Lubbock's 1871 Bank Holiday Act as well as cheaper rail fares, seaside resorts like Brighton had become accessible to the lower-middle and upper-working classes (the so-called aristocracy of labour). Thanks to overzealous collecting of specimens and visitors' sewage,

[16]Harvard University, Cambridge, MA. Harvard Theater Collection, MS Thr 419, ser. II.

these resorts no longer afforded the beachcomber much by way of wild specimens. But that was no hardship, when one could admire the 'wonders of the shore' in tanks, without getting one's feet wet, and laugh at the clowns.

Back in London, the Royal Aquarium took this process to its logical conclusion. Although it opened in 1876 with a highly engineered system of supplying mammoth tanks with fresh- and seawater, the fish never came. A dead whale was displayed in 1877, but otherwise the tanks were used for displays in which female athletes demonstrated different swimming strokes. Although these taught a useful, potentially life-saving skill, one suspects that it was the unfamiliar sight of attractive females in wet, skin-tight swimsuits which formed the main attraction. Other fragrant females such as 'Zazel' were shot out of cannons, an act supposedly invented by the American tight-rope-artist and Aquarium impresario William Hunt ('The Great Farini').

Farini stage-managed the display of a child known as 'Krao' at the Aquarium in 1887. Supposedly 'captured' by the Norwegian Carl Bock in northern Laos, 'Krao' was a human child with hypertrichosis (large amounts of hair, covering the entire body). On the Aquarium's promotional material, however, she appeared as 'The Missing Link' (Fig 6). Farini's manipulation of the media and the scientific press, including the *British Medical Journal* and *Science*, managed to transform what was essentially a 'bearded lady' freak show act into an important scientific discovery. The adventure story of her 'capture' by Bock, a tale of derring-do against the odds added another layer of interest. Laos was then on the fringes of the British Empire, a supposedly 'lawless' realm of 'savages'. Farini played on his audience's fascination for the few 'uncivilized' places remaining between the territories of European powers and their settler colonies. 'Krao' not only proved Darwin right, but also confirmed Britons' view of empire as a progressive (and therefore 'evolutionary') process of civilization.

Treasuring and teaching

In 1881 the Natural History Museum opened in South Kensington, one among a complex of national institutions intended to serve the arts and sciences, built on land purchased with the profits of the 1851 Great Exhibition. Although it remained a branch of the BM, the transfer of the latter's natural history collections from its cramped home in central London to a massive purpose-built museum drew a line under the Enlightenment idea of the museum as one embracing both *naturalia* and *artificalia* (i.e. man-made things). The NHM was designed by Alfred Waterhouse in a Romanesque style associated with the churches of twelfth-century Germany. Its main entrance resembles that of a medieval cathedral.

The original competition to find an architect had been held in 1862. The winner, engineer Francis Fowke, died 3 years later, creating an opening for Waterhouse. Construction was severely delayed by the thriftiness of

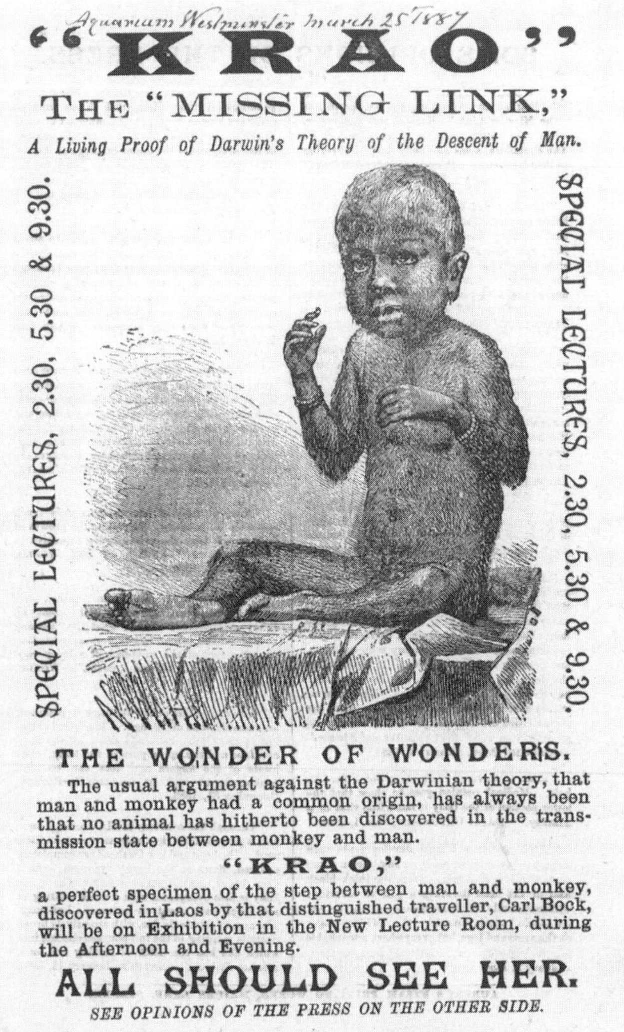

FIGURE 6 *Playbill for the Royal Aquarium, London, 1887. Courtesy of the City of Westminster Archives Centre.*

Gladstone's ministers, but also by debates over how such an institution was to be laid out and how 'architectural' it should be. As Ruskin had noted, all museums had not only a 'treasuring' but also a 'teaching' role, and those roles were performed in front of different audiences. They were treasure houses or repositories housing large numbers of specimens, to which experts could have reference in the course of their research. But they also had a responsibility to teach basic principles to a much larger, non-expert audience. Instead of expecting them to make sense of the vast collections confronting them, it was far better to select a few specimens which would capture the attention and illustrate abstract principles.[17]

Beginning in the 1830s, the so-called Gothic Revival had seen British architects turn away from the Classical styles (i.e. those of Antiquity, of ancient Greece and Rome) in which they had been building since at least the seventeenth century. The ancient Greeks, it was observed, had not built their temples to suit the British climate or to honour a Christian God. The pointed-arch Gothic style was more suited to a Christian nation and to a race like the British, descended from forest-dwelling Teutons (Gothic vaulting supposedly resembled tree canopies). New materials (concrete, steel girders, cheap glass), however, had brought about a separation between style and construction. On the one hand Victorians could achieve unprecedented technical feats such as the 'Crystal Palace': a glass-and-steel cathedral erected in months, whereas the builders of medieval stone-built cathedrals had taken decades. On the other hand, Victorians struggled to develop a new style of architecture they could call their own.

In terms of the actual building of a natural history museum, therefore, it was unclear whether this represented an engineering or an architectural problem. Was a natural history museum building simply a container or shell (an engineering challenge, of the kind Fowke was used to tackling), or could an architecture (a style, a language of ornament) be found, which 'spoke to' the objects and ideas it housed? The Oxford University Museum of Natural History, in which the famous Wilberforce/Huxley exchange had occurred, was an important precedent. Opened in 1860, it had been designed by T. N. Deane and Benjamin Woodward in a Gothic style, that is, one that also looked back to the Middle Ages. Its lecture theatre was a copy of the kitchens of a ruined monastery in Glastonbury, Somerset. Its courtyard was a forest of steel columns supporting a glass roof; Gothic forms rendered in new materials.

Walking in the Oxford Museum's Library, the architectural historian James Fergusson felt the medievalizing design to be dishonest, a fake. 'You take a book from [the shelf] and are astonished to find that men who could spend thousands on thousands on this great forgery have not reprinted Lyell's "Geology," or Darwin's "Origin of Species" in black letter, and illuminated them, like the building, in the style of the thirteenth century.'[18] Natural history could not, it seemed, 'speak to' Gothic Revival architecture, because the Gothic was the product of an age of superstition. Owen initially agreed. 'The Sciences were not born nor nursed,' he noted in 1855, when 'that style originated.'[19] The architect George Edmund Street however argued that the Gothic was the natural choice in a building 'intended

[17]*Select Committee on Public Institutions* (Parliamentary Papers, 1861, 16), Q1596.

[18]James Fergusson, *A History of the Modern Styles of Architecture: Being a Sequel to the Handbook of Architecture* (London: John Murray, 1862), p. 328.

[19]Cited in Carla Yanni, *Nature's Museums. Victorian Science and the Architecture of Display* (London: Athlone, 1999), p. 89.

mainly for the reception of a collection illustrative of Natural History'. Starting in the 1840s Gothic Revivalists like Street and George Gilbert Scott celebrated Gothic architecture as an architecture of 'development', that is, in evolutionary terms.[20]

The NHM is interesting to us not because it reached a solution of these problems, but because it represents something of a fudge. The main hall of Waterhouse's museum feels like a cathedral nave, the side-compartments like small chapels. It also feels like a Victorian train shed: the ceiling is partly glass, and the girders supporting the main staircase are gaily painted, rather than hidden. Terracotta monkeys are shown playfully climbing the arch of the main vault, but this opportunity to show an evolutionary lineage is neglected. Outside, the main front was surmounted by a figure of Adam, reflecting the privileged position of *Homo sapiens* in the Genesis story.

Owen's original 1859 sketch for the museum proposed that this nave would lead on to an Index Museum, a systematic compendium of all British flora and fauna. After toying with a lecture theatre, however, this plan was dropped. Owen also proposed, then dropped plans for an introductory display of ethnology in the main hall. Plans to put 'teaching' spaces at the core of the NHM were left in suspense, therefore. Instead the Hall has become a kind of valhalla or shrine housing statues of great natural historians, notably Owen, Darwin and Huxley. The relative positioning of these statues (and the absence of Wallace) continues to be 'read' as a virtual history of British natural history, rather than anything to do with natural history itself.

A happier solution was proposed in the parallel galleries which run off the main corridor of the NHM, which bisects the Hall. Here wide, top-lit double-height galleries are divided by lower, narrower galleries. The former were intended for the public, with a few 'teaching' specimens well spaced out in display cases on either side. The narrow galleries were intended for the curators, with spaces for experts to study the specimens they had selected from the thousands stored in nearby high-density shelving units. Some plans even suggest that the vitrines would be double-sided, so that the public could look through them to the specialists working inside, observing scientists at work as one might observe gorillas at the ZSL.[21] Although this scheme was never fully realized in the Victorian period, the concept of 'transparency', of displaying the museum's 'hidden' research activity to casual visitors is now considered cutting edge. At the NHM's Darwin Centre Cocoon visitors can look through windows into the curators' offices and even speak to them through a microphone.

The balance between 'treasuring' and 'teaching' achieved at the Victorian NHM was provisional, therefore. The lessons taught were not evolutionary,

---

[20]G. Alex Bremner and Jonathan Conlin, 'History as Form: Architecture and Liberal Anglican Thought in E. A. Freeman,' *Modern Intellectual History* 8(2) (2011): 299–326.
[21]Yanni, *Nature's Museums*, pp. 127–30.

but classificatory, as an expression of the ordered diversity of the Creator's mind. Although there was much more space at South Kensington, otherwise Grant's criticisms of the old BM's displays made during the 1835 Select Committee (see Chapter 2) remained valid. Like the Bridgewater Treatises or the popular handbooks discussed here, such natural history museums sought to instil humility and wonder at God's creative intelligence. The leading mind behind the Oxford Museum, professor of Medicine Henry Acland co-opted evolution when he wrote to Owen in 1862 claiming that 'Whatever views Mr Huxley, or you, or Mr Darwin, or the Bishop of Oxford may have as to the essential Nature of Man, you all agree that however he so became, he is in some manner made . . . by the ordinance of God.'[22]

The variety of publications, performances and displays described here suggests a sustained desire on the part of a wide upper working- and middle-class public to be informed on what were clearly felt to be important 'issues of the day'. With the widening of audiences in the 1870s came a shift in understanding of what 'science' was. It did not represent a fixed body of data or theories to be learnt, but a form of discourse, a part of public life which the responsible citizen ought to be conversant in, ought to 'keep up with'. This model of science, as a variety of cultural literacy, could be profoundly unsettling to those educators, notably the schools inspector Matthew Arnold, author of *Culture and Anarchy* (1869), who insisted on the primacy of a cultural tradition founded on Antiquity. Two decades later Huxley was equally concerned, except he claimed that scientific conversation had collapsed into polite chatter about Theosophy, spiritualism and other topics he saw as little more than fads.[23] Science was followed more as a source of titillating speculations than as an attempt to establish a unified theory of the universe.

Further reading

Durbach, Nadja, *Spectacles of Deformity. Freak Shows and Modern British Culture* (Berkeley: University of California Press, 2010), ch. 3.

Lightman, Bernard, *Victorian Popularizers of Science. Designing Nature for New Audiences* (Chicago: University of Chicago Press, 2007).

Lightman, Bernard and Aileen Fyfe (eds), *Science in the Marketplace. Nineteenth-century Sites and Experiences* (Chicago: University of Chicago Press, 2007).

O'Connor, Ralph, *The Earth on Show. Fossils and the Poetics of Popular Science, 1802–1856* (Chicago: University of Chicago Press, 2007).

Paradis, James, 'Satire and Science in Victorian Culture,' in Lightman, Bernard (ed.), *Victorian Science in Context* (Chicago: University of Chicago Press, 1997), pp. 143–75.

---

[22]Cited in Ibid., p. 89.

[23]James Secord, 'How Scientific Conversation Became Shop Talk,' in Aileen Fyfe and Bernard Lightman (eds), *Science in the Marketplace* (Chicago: University of Chicago Press, 2007), p. 46.

Secord, Anne, 'Science in the Pub: Artisan Botanists in Early Nineteenth-century Lancashire,' *History of Science* 32 (1994): 269–315.

Secord, Jim, 'Monsters at the Crystal Palace,' in Soaraya de Chadarevian and Nick Hopwood (eds), *Models: The Third Dimension of Science* (Stanford: Stanford University Press, 2004), pp. 138–69.

Vincent, David, *Bread, Knowledge and Freedom: A Study of Nineteenth-century Working-Class Autobiography* (London: Methuen, 1981).

Yanni, Carla, *Nature's Museums. Victorian Science and the Architecture of Display* (London: Athlone, 1999).

Zuckerman, Solly, Baron Zuckerman, *The Zoological Society of London* (London: ZSL, 1976).

CHAPTER NINE

Sustainable evolution? Alfred Russel Wallace and the Wonderful Century

FIGURE 7 *William Strang,* Alfred Russel Wallace, *1908. One of a series of portrait drawings commissioned by Edward VII to record members of the Order of Merit, an order of chivalry he established in 1902. For all his involvement in controversial causes, Wallace nonetheless collected many such honours in later life.*

In 1909 the *Daily Mail* invited Alfred Russel Wallace to write an appreciation of Darwin, to mark the centenary of the latter's birth, and the 50th anniversary of *The Origin*. Like Huxley in 1880, so Wallace exaggerated somewhat to make a point, claiming that belief in the fixity of species had been 'almost universal' before 1859. Though Huxley had modestly referred to himself as 'undernurse' to *The Origin*, Wallace did not refer to his own role or discovery at all. Together with *The Descent of Man* Darwin had convinced readers that they and the apes had evolved from a common ancestor by natural selection:

> In this conclusion the great majority of thinkers to-day are in entire agreement with [Darwin]; but in his further contention–that the whole mental, moral, and aesthetic nature of man has also descended to him from the lower animals–a large and increasing number of his admirers do not follow him, there being at least as much positive evidence against as in favour of his contention.[1]

As we have seen, Huxley's celebration of *The Origin*'s 21st birthday had contained a warning: having started life as a 'heresy', it might finish as a 'superstition'. Wallace's celebration also strikes an ambivalent note. Surely by 1909 everyone recognized that humans inherited their mental faculties and social instincts through the same process by which they had inherited their skeletal structure, brain and other organs? Where else could they have come from?

Wallace's relationship with Darwin and his theories of natural and sexual selection is even more curious than that of Huxley. Though the similarities in terminology and content between Wallace's theory and his own had shocked Darwin when he received Wallace's letter in 1858, there were differences in emphasis. Whereas Darwin focused on competition among individuals, Wallace thought more in terms of population. Adaptation was a means of sustaining more life as much as it was about 'fitter' life. Wallace's interest in the distribution of life in time and space kept him in Malaya for 4 years after 1858, refining what we know today as 'Wallace's Line'. Books such as *The Geographical Distribution of Animals* (1876) and *Island Life* (1880) laid the foundations of biogeography.

Like Darwin, Wallace had cut his teeth collecting and classifying beetles. His very different social background and two solo expeditions, however, gave Wallace prolonged experience of working-class life in industrializing Britain as well as of 'savage' life beyond the edges of any empire. In meeting the socialist Robert Owen at his 'Hall of Science' in London in 1837 Wallace had also encountered an alternative to free-market capitalism, beginning an interest in political economy which Darwin did not share.

---

[1]Wallace, 'To-day's Centenaries: Charles Darwin.' The Alfred Russel Wallace Page, S672. http://people.wku.edu/charles.smith/wallace/S672.htm (accessed 2 May 2013).

When it came to defining human 'fitness' and human 'progress' Wallace was able to see several possible criteria, several different alternatives, where Darwin could only see one.

In 1865 Wallace attended his first seance, that is, a meeting at which a spiritualist medium (usually a woman) went into a trance and passed messages of different kinds between 'the spirit realm' and a group of individuals gathered around a table, sometimes in the dark. Within a year he was in print urging his fellow men of science to engage with spiritualism. Within 4 years he had concluded that natural selection alone could explain neither the origin of humanity's moral and intellectual faculties nor certain morphological features. Though Darwin perceived this as betrayal of their 'child', Wallace saw a role for natural selection within his 'universe of spirit'. Wallace felt Darwin had diluted natural selection in coming up with sexual selection, a theory Wallace never accepted, as well as in his readiness to allow Lamarckian and other odds-and-ends to slip back into later editions of *The Origin*. Wallace defended natural selection long after Darwin died, while others were seduced by Weismann's 'germ plasm' and Mendelianism.

Spencer let Wallace down in a similar fashion. Though Spencer was 2 years younger, Wallace revered him and his *Social Statics*, naming his first son Herbert Spencer Wallace. Wallace's campaign against compulsory state vaccination reflected a Spencerian suspicion of the state. In 1881 Wallace became president of a new Land Nationalisation Society. Unfortunately Spencer refused to be associated with it, despite having advocated nationalization in *Social Statics*, arguing that such a change could not be realized as a deliberate project, let alone a political one. Wallace nonetheless continued to argue for radical reform of the relationship between capital and labour. In 1889 he proclaimed himself a socialist.

Kingsley, Darwin, Huxley, Spencer: Wallace outlived them all, dying in 1913, in his 90th year. Longevity secured him the privilege of surveying the nineteenth century from an evolutionist's perspective, resulting in *The Wonderful Century: Its Successes and Failures* (1898), which, as its title suggests, endeavoured to offer a balanced assessment of Victorian progress. Yet the length of his career has led some biographers to represent it as made up of distinct phases or campaigns which were at odds with each other. This book may unwittingly have encouraged such a view by considering Wallace's career up to 1865 in an earlier chapter (Chapter 3).

The historian of science Robert M. Young argued that socialism and 'evolutionism' were 'very uneasy bedfellows', and that Wallace chose the former over the latter.[2] Yet socialism was far from being a well-established or clearly defined ideology in Wallace's day. Although Karl Marx, Friedrich Engels and Vladimir Lenin all spent time in exile in London during our period, they had little impact outside a very narrow circle. Wallace's 1889

---

[2]Robert M. Young, '"Non-scientific" factors in the Darwinian debates,' *Actes du XIIe Congrès Internationale d'Histoire des Sciences*, 12 vols (Paris: Albert Blanchard, 1971), 8: 221–6 (224).

declaration has something of Kingsley's earlier self-identification as a Chartist about it; a case of taking on a scandalous label in a deliberate attempt to challenge familiar stereotypes of what 'a Chartist' or 'a socialist' looked and sounded like. There may be a similar sense of an outsider trying to 'catch up' with a movement in order to shift its direction.

Three years before Karl Marx's *Das Kapital* (originally published in German in 1867) had made its first appearance in English. Its impact on British socialism was far less profound, however, than the Bloody Sunday riots, which took place around London's Trafalgar Square in November of that same year. Police and special constables used violence to scatter a peaceful march by unemployed workers, symbolically denying them access to the public square to make their voice heard. The march raised, then dashed the hopes of Annie Besant, William Morris and other members of the Socialist League, the Fabian Society and other socialist societies that a transition from capitalism to socialism could be brought about in Britain. In the aftermath some, like Morris, withdrew into quasi-utopian cooperatives and other projects that looked back to a very English vision of peasant pastoralism. Besant left Britain for India in pursuit of mystic teachings. The Fabians buried themselves in sociological analysis and patiently awaited the day when the rise of labour as a force within the British polity would give them the opportunity to enact radical social change through central government.

Though they differed in the extent to which they looked to the state or the British people to bring about change, Victorian socialists, communists and anarchists all looked to the natural world and to evolutionary processes for justification of their beliefs. Apart from his symbolic role as a giant of Victorian natural history who also called himself a socialist, Wallace's contribution to the socialist movement lay in his critique of capitalism (and by extension empire) as an unsustainable means of allocating finite natural resources. It may not be necessary to put Wallace's socialism or even his spiritualism to one side, as embarrassing wrong-turns. It may be possible to represent his 'spirit world' as an internally consistent model, one established by God to foster diversity, of species, but also of ideas. From the debate over spontaneous generation to that over extra-terrestrial life, Wallace was fascinated by the possibilities of life. Seen in this way evolution becomes a process of increasing the universe's sustainability, that is, its ability to maintain life, in as many forms as possible.

Spiritualist science

Wallace's first contact with spiritualism came in the early 1860s, via Robert Owen. Owen's son, Robert Dale Owen, had written a book entitled *Footfalls on the Boundary of Another World* (1859), in which he described the dead haunting living animals and humans, apparitions of living people, clairvoyance and automatic writing (i.e. texts generated by spirits, either directly or through a medium). Such ideas built on ideas of animal magnetism popular in Enlightenment circles in the previous century. The German astronomer Franz Mesmer gave his name to one form of hypnosis, one Wallace successfully attempted on his school pupils in the 1840s, and later on Malayans. Indeed, Wallace's first publication was an 1845 letter describing the former experiments published in *The Critic*. But Mesmer does not seem to have believed that hypnosis could be used to communicate with the dead. The eighteenth-century Swedish mystic Emanuel Swedenborg did and posited the existence of a number of planes of existence. Swedenborg's New Church (the Swedenborgians) had a very small but select following in Victorian Britain.

The first seance Wallace attended was organized by his sister Fanny, and the medium was her housekeeper. The Wallaces' declining financial fortunes had led Fanny to emigrate to the United States and work as a teacher. The spiritualist movement had taken off in the United States in the 1850s, with various mediums claiming to have communicated with the spirits of the dead by voice and writing as well as through poltergeist-style rappings. Although there were moves to build a set of doctrines and even a church on such manifestations, for the most part the 'movement' was diffuse and slow to organize. The Society for Psychical Research (SPR) and the College of Psychic Studies in London are two societies with distinguished Victorian origins, established in 1882 and 1884, respectively. Wallace was somewhat nervous of holding office in such groups. Shy in society, Wallace rightly felt the best way he could serve was by writing: books, pamphlets and letters. Wallace constantly bombarded the editor of the *Times* and other newspapers with letters. On several occasions he gave evidence in court in defence of mediums accused of infringing the Vagrancy Act.

Wallace's position as a latecomer to spiritualism on the fringes of the movement contrasted with his theoretical work stitching together natural selection and spiritualism and with his attempts to encourage experimental research into the psychic forces at work during manifestations. In his 1869 review of the 10th edition of Charles Lyell's *Principles of Geology* Wallace praised Lyell for the 'youthfulness of mind' evinced by his willingness to accept natural selection and so abandon 'opinions so long held and so powerfully advocated'. Wallace was arguing that Lyell's colleagues should show the same spirit with regard to spiritualist phenomena. They needed to keep an open mind because neither natural selection 'nor the more general theory of evolution' could explain how life originated.

As we have seen, Darwin, Huxley and others were deeply uninterested in this question. The physiologist Henry Bastian (1837–1915) became active in this area around the same time as Wallace became interested in spiritualism. His book *The Beginnings of Life* (1872) argued that Darwin's theories presumed that spontaneous generation took place. Huxley's response was to lead a campaign to discredit Bastian and his work. Active between 1864 and 1892, the nine-member X-Club whose monthly meetings brought Huxley together with other rising and influential stars in the world of science facilitated such attempts to police the borders of natural history. Huxley considered Bastian a sloppy experimental scientist, but he also had political reasons, concerned not to yoke his scientific naturalism with either materialism or the randomness implied by the adjective 'spontaneous'. Though Darwin privately conceded that such generation (alternatively known as archebiosis or abiogenesis) was likely, in public Huxley argued that, if abiogenesis had ever occurred, instances had been restricted to the distant past.[3]

Wallace's 1869 review also identified several morphological features which could not be explained by natural selection, because they exceeded the requirements of primitive humans or hominids. Why did 'savages' have a brain so much larger than that of an ape, when they only had need of one slightly larger? *Homo sapiens'* upright posture, hand, the arrangement of organs used to produce speech, all represented cases of 'an instrument' having been 'prepared in advance of the needs of its possessor'. Paley had referred to similar 'anticipatory' organs in his *Natural Theology*. Yet whereas Paley had refused to accept transmutation Wallace argued that one could accept the laws of natural selection and other 'laws of organic development' as explaining the origin of humanity and all life *and* acknowledge these traits and faculties as 'evidence of a Power which has guided the action of those laws in definite directions and for special ends'.

It was at this point that Darwin wrote 'No' in the margin of his copy.[4] Wallace went on to turn Darwin's analogy between natural and artificial selection against the latter. Wheat and the London dray horse were both products of artificial selection with a strong resemblance to other life forms which had not been bred by humans. A human who understood the 'laws of organic development' might refuse to acknowledge that a 'Higher Power' had played a role in guiding their development. Yet we knew that humans had played such a role, and so, Wallace concluded, we must admit the possibility that an even higher power had guided those laws for a 'nobler aim' in the case of humans.

> Such, we believe, is the direction in which we shall find the true reconciliation of Science with Theology on this most momentous problem. Let us fearlessly admit that the mind of man (itself the living proof of a supreme mind) is able to trace, and to a considerable extent has traced, the laws by means of which the organic no less than the inorganic world has been developed. But let us not shut our eyes to the evidence that an Overruling Intelligence has watched over the action of those laws, so

[3]James E. Strick, *Sparks of Life. Darwinism and the Victorian Debates over Spontaneous Generation* (Cambridge, MA: Harvard University Press, 2000), ch. 4.
[4]Adrian Desmond and James Moore, *Darwin* (London: Michael Joseph, 1991), p. 569.

> directing variations and so determining their accumulation, as finally to produce an organization sufficiently perfect to admit of, and even to aid in, the indefinite advancement of our mental and moral nature.[5]

In his *Footfalls* Dale Owen had devoted several pages to discussing the precise meanings of 'impossible', 'improbable' and 'miraculous'. Wallace followed suit at greater length in his *The Scientific Aspect of the Supernatural* (1866), a work later revised as *On Miracles and Modern Spiritualism* (1875). Theologians and natural historians had long been exercised by the miraculous. Was it possible for God to interrupt the action of the secondary laws, to contravene the rules of matter and motion which He himself had established? Even assuming He could do so, would this not be somehow demeaning, suggesting that there was some error that had crept into Creation (from where?). In this way something which could be dismissed as an American fad could appear part of a distinguished tradition of metaphysical discussion. As the introduction indicated, these questions of the authority of scientific facts and hypotheses are fundamental to science.

Wallace pointed out several ways in which the 'supernatural' was a relative rather than absolute term, contingent on the observer's prior knowledge or the observer's position in history. A century ago, he noted, information passed from a distant continent by telegraph would have been considered miraculous, evidence of clairvoyance, perhaps. Gravity was a fundamental force without which physics and engineering would be impossible. Magnetism and electricity were other forces the nature of which was unobservable except indirectly, by studying how it interacted with matter. Yet these were invisible forces, acting at a distance. How much of a leap was it, therefore, to imagine spirits acting upon physical objects?

That 'science of energy' practised by Tyndall, William Thomson (Lord Kelvin) and others was leading us to a new understanding of how different forms of energy (heat and light, say) related to one another. 'Light, heat, electricity, magnetism, and probably vitality and gravitation, are believed to be but "modes of motion" of a space-filling ether,' Wallace noted, 'and there is not a single manifestation of force or development of beauty, but is derived from one or other of these.' How much of a leap was it, therefore, to imagine spirits interacting directly with these invisible forces? While our human senses only revealed part of them to us, that need not be the case of 'beings of an etherial order', who might just as easily have a Higher Intelligence as well. If those 'beings' were to exert their powers in a way humans could perceive, 'the result would not be a miracle, in the sense in which the term is used by Hume or Tyndall'. 'There would be no "violation of a law of nature"'; Wallace insisted, 'there would be no "invasion of the law of conservation of energy."'[6]

---

[5]Wallace, 'Sir Charles Lyell on Geological Climates,' *Quarterly Review* April 1869. The Alfred Russel Wallace Page, S146. http://people.wku.edu/charles.smith/wallace/S146.htm (accessed 2 May 2013).

[6]Wallace, *The Scientific Aspect of the Supernatural* (London: F. Farrah, 1866), pp. 8–9.

Although important as a theorist of spiritualist science, Wallace himself did little by way of applied spiritual or psychical research beyond collecting the names of influential individuals who believed and accounts of spiritualist manifestations. A field biologist, he may simply have lacked the skills in physics necessary to conceive experiments that might yield useful data. Wallace did what he could: he noted that spirits obeyed Wallace's Line in his *Malay Archipelago* (ghosts were believed to exist on one side of the line, but not the other) and when a bouquet of flowers materialized during a seance, he identified what species of flower they were.[7]

An expert in photography and spectroscopy who had discovered the element thallium in 1861, the chemist and physicist William Crookes (1831–1919) was much better placed than Wallace to investigate. In Daniel Douglas Home and, later, Florence Cook, Crookes found mediums willing to cooperate. In Home's case Crookes found evidence that he could control gravity using psychic force and produce musical effects by the same means. Though elected a Fellow of the Society back in 1863 for his work on thallium, the Royal Society refused to publish his findings. Crookes published them in his own *Quarterly Journal of Science*. The British Association for the Advancement of Science (BAAS), too, refused to allow him to give a paper on spiritualist phenomena. This frustrated Wallace, who made a point of including a paper on clairvoyance in the Anthropology section in 1876, when he was chairing that section of the BAAS.

Crookes continued his research fitfully thereafter. After experiments with the medium Mary Showers conducted in 1874–75 unearthed evidence of deception he ceased to collaborate with mediums. He found evidence of psychic force, however, using a radiometer, the 'lightmill' device he had invented in 1873. Though the vanes of a Crookes' radiometer do turn when a hand is nearby, this is the result of radiation (the heat given off), not psychic force. These experiments led Crookes to discover cathode rays and invent the cathode ray tube, the technology that made the discovery of the electron (and television) possible.

In contrast to Wallace and Crookes, many men of science attended seances in an aggressively sceptical rather than curious or receptive spirit. Though Darwin had spiritualists in the family (his cousins Francis Galton and Hensleigh Wedgwood; his own son, George) he refused invitations to attend seances and left the one he did attend halfway through. 'It is rather dreadful,' he observed, 'to think what we may have to believe.'[8] The quote suggests a categorical, unscientific refusal to test his own beliefs. When Darwin was asked by Asa Gray what evidence he would accept as disproving evolution by natural selection, Darwin could not think of a reply, beyond a rather feeble joke (if an angel came and told him so, he proposed).

The question was far from being a joke. Indeed, it anticipates the test of 'falsifiability' later proposed by the philosopher of science Karl Popper.

---

[7]Wallace, *The Malay Archipelago: The Land of the Orang-utan, and the Bird of Paradise* (New York: Harper, 1869), p. 124.

[8]Ross A. Slotten, *The Heretic in Darwin's Court: The Life of Alfred Russel Wallace* (New York: Columbia, 2004), p. 315.

Popper posited that a hypothesis was valid only if it was possible to imagine something that would prove it to be false. Darwin's response to Gray's question implies that evolution is not falsifiable. Happily, where Darwin failed, others have since succeeded: J. B. S. Haldane noted that finding fossil rabbits in Cambrian rock would falsify the theory. Taken together with his rejection of spiritualist enquiry, however, it suggests that Darwin may have lost his 'youthful mind' as he grew older.

Wallace's book *Darwinism* (1889) continued to posit a three-stage universal evolution of life, each involving different degrees of 'spiritual influx'. The first stage saw a shift in gear from 'protoplasm as a chemical compound' to '*living* protoplasm'. The second stage saw the emergence of consciousness, the third that of intellect (in humans). Taken together they pointed 'to a world of spirit, to which the world of matter is totally subordinate'. 'To this spiritual world,' Wallace continued, 'we may refer the marvellously complex forces which we know as gravitation, cohesion, chemical force, radiant force, and electricity, without which the material universe could not exist for a moment in its present form, and perhaps not at all.'[9]

In the 1890s Wallace became less interested in psychical phenomena, and he played little role in the work of the SPR, which took its momentum from 'The Souls', a close-knit and snobbish aristocratic coterie of 'bright young things' whose manners and culture were totally alien to Wallace. Under Frederic William Henry Myers (1843–1901) it developed a theory of mind divided between a self-aware or 'supraliminal' element and a 'subliminal' unconscious (from the Latin for threshold: limen). Myers was among the first to introduce the ideas of the Austrian Sigmund Freud to Britain and was influenced by Freud's understanding of the subliminal as a stage for repressed 'lower' drives. Repression and atavism were unappealing to Wallace, who saw the spirit world as one of organic unity, purity and moral refinement. Wallace probably saw Freud as an irrational if nonetheless fashionable foreign fad, similar to the Theosophy peddled in Britain by the mysterious Madame Blavatsky around 1880. Though it appealed to many spiritualists, for Wallace Theosophy's mix of Zoroastrianism, Buddhism and Christianity was 'purely imaginative', delighting in obscure ritualism for its own sake.[10]

The enthusiasm with which 1890s intellectuals (and wannabe intellectuals) toyed with spiritualism, Theosophy and Spencer's 'Unknowable' nonetheless served to make Wallace's questing scepticism seem more exciting than the increasingly scientistic model of 'professional science' championed by Huxley. Huxley's unsuccessful attempts to reclaim 'agnosticism' for his variety of natural science reflect this decadent 1890s climate of doubt and credulity, irony and sentimentality familiar from Oscar Wilde.

---

[9]Alfred Russel Wallace, *Darwinism: An Exposition of the Theory of Natural Selection with some of its Applications* (London: Macmillan, 1889), p. 476.

[10]Cited in Martin Fichman, *An Elusive Victorian: The Evolution of Alfred Russel Wallace* (Chicago: University of Chicago Press, 2004), p. 302.

FOUNDATIONS OF AGNOSTICISM

Among Myers' associates in 'The Souls' was Arthur James Balfour (1848–1930), nephew of the 3rd Marquess of Salisbury, from whom he took over as Conservative Prime Minister in 1902. Although Balfour entered Parliament in 1874 and held cabinet office under his uncle in the 1880s and 1890s he retained a lively interest in philosophy, spiritualism and the natural sciences. His brother was an embryologist, the physicist John Strutt a close friend. In 1894 his uncle had served as president of the BAAS, delivering a Presidential Address entitled 'Evolution: a retrospect' whose survey of the authority and achievements of the natural sciences was as sceptical and hedging as Tyndall's Belfast address of 1874 had been assertive. The best one could conclude, Salisbury claimed, was that natural selection was 'not proven'.

In 1895 Balfour followed up with his *Foundations of Belief*. In it he noted the pretensions of 'the naturalistic hypothesis' to explain moral sentiments and beliefs in evolutionary terms, as 'merely samples of the complicated contrivances, many of them mean and many of them disgusting, wrought into the physical or into the social organism by the shaping forces of selection and elimination'. The 'oracle' of 'naturalism' had spoken, but the answers were 'unsatisfactory': our virtue was mere instinct, our world just a speck in the void. Unable to 'swallow this strange universe' of their own creation, Huxley and Spencer had denied its existence as external reality. The 'consistent agnostic' was left with nothing to do but sit and contemplate 'the long procession of his sensations', without bothering with pointless investigation of what might lie beyond.[11] Modern humans, in short, had to make a choice between 'naturalism' and a theism more respectful of our soul, reason and beauty.

Salisbury's Presidential Address had left Spencer fuming, urging Wallace to once again take up the cudgels on behalf of natural selection. Though Wallace refused, citing ill health, Huxley did fight back against Balfour in *The Nineteenth Century* in March 1895. He titled his response 'Mr Balfour's Attack on Agnosticism', as well he might, considering that Huxley had coined the word 'agnostic' back in 1869. *Gnosis* is the Greek word for 'knowledge' and so Huxley's identification of himself as a 'non-knower' or sceptic was designed to protect himself from charges of materialism as well as to distance himself from Spencer's Unknowable-with-a-capital-'U'. Unfortunately Huxley failed to assert his understanding of the term, and by the time he did so (1889), his insistence that agnosticism was a method, not a cover for positivism (as Balfour thought), was unconvincing to many.

Huxley's response was tragically ineffective, and not just because he died before the second part of 'Mr Balfour's Attack' could be published. In his response Huxley constructed a safely Anglican pedigree for his 'agnosticism', while he challenged Balfour to find any basis for his 'theologism' in Christian doctrine. Though the

[11] A. J. Balfour, *A Defence of Philosophic Doubt: Being an Essay on the Foundations of Belief* (London: Longmans, 1906), pp. 21, 86, 134.

knowledge of church history deployed by Huxley was impressive (particularly for an 'atheist'), Huxley's presumption that Balfour (or other readers of *The Nineteenth Century*) still cared about Christian 'orthodoxy', about the pedigree of this or that belief, was wide of the mark. They wanted something to believe in, and, thanks to thinkers like William James, no longer saw this 'will to belief' as a flaw. They certainly did not associate it with that cynical, political courting of 'opinion' which Huxley called 'demomism' (and associated with Gladstone).

For them Huxley's insistence that 'the physical powers . . . determine the psychical' was determinist; his conclusion that 'the correlating bond' linking them was 'an insoluble riddle' presumptuous and tragic at the same time.[12] Huxley also presumed a clear border between knowledge and ignorance, at a time when many preferred the fuzzier concept of the 'probability of evidence' originally advanced by the Anglican theologian Joseph Butler in his *Analogy of Reason* (1736). As Balfour later intimated in *Theism and Humanism* (1915), Huxley's insistence on both the 'hardness' and fixity of this border seemed to deny what should have been obvious to an evolutionist: that nothing, including Huxley's own scientific methods, was exempt from change.

Land and labour

Wallace seems to have been inspired to consider questions of political economy by observing the efforts of the industrialist and socialist Robert Owen to model a new, collectivist form of manufacturing community at New Lanark and at New Harmony, a model community established in Indiana in 1824. These communities were intended to be self-supporting islands of cooperation, built on land bought and held in common. In the long term it was hoped that they would thrive, establishing new offshoots and using their surplus to acquire ever more land. By tapping humanity's social passions and natural altruism a revolution in human society could be brought about, peacefully and without having to get bogged down in party politics. By a kind of 'survival of the fittest' cooperative communities would outbreed, expand and eventually replace those founded on individualist, capitalist competition.

Unfortunately Owen's advocacy of easier divorce and other reforms to traditional marriage were widely misinterpreted as support for free love. Owenite communities attracted a motley collection of elite idealists and independent-minded artisans; it proved a struggle to find the right balance of enthusiasm, tolerance and practical skills. Wallace came to recognize the need for targeted recruitment to secure a better balance, as well as the challenges involved in integrating individuals who had been brought up

---

[12]Huxley, 'Mr. Balfour's Attack on Agnosticism – II.' The Huxley File, http://aleph0.clarku.edu/huxley/Mss/Balfour2.html (accessed 2 May 2013).

with a very different set of guiding principles. Although today historians of political thought struggle to integrate Owen into histories of socialism, in his autobiography Wallace described Owen as 'the greatest of social reformers and the real founder of modern Socialism'.[13] Together with his enthusiastic reading of *Social Statics* (1850), Owen encouraged Wallace to see socialism as a recipe for practical social reform, not a socialist politics.

Although Spencer's book did inspire the ending to *The Malaya Archipelago* (1869), with its rant against his fellow Britons' supposed 'civilization', otherwise Wallace did not address questions of land, capital and labour in print until the 1880s. During the 1870s, however, attempts to address unrest in Ireland as well as the effects of agricultural depression at home brought the rights of aristocratic landowners and their tenants to the fore. For centuries aristocratic families had used primogeniture (the system by which the eldest son inherits, rather than sharing their deceased relative's estate with siblings) and entail to maintain the integrity of their family's estates, preventing this or that generation from selling land. New legislation helped such families to break entail, releasing land onto the market. Under Gladstone and later, Salisbury, the rights of Irish tenants were extended, giving them longer, more secure tenures and better compensation for any improvements they made.

In Ireland changes to tenant-right were intended to create a new class of independent proprietors, replacing the older system of vast aristocratic estates, many of them controlled by English aristocrats, who rarely visited Ireland. These measures were more about political expediency than anything else. By creating a more balanced and therefore cohesive Irish society, land reform would, it was hoped, take the wind out of the sails of Irish Home Rule and remove the destabilizing effects of Irish MPs on British politics. In the 1880s the Fenians began staging terrorist attacks on English railway stations, further increasing the pressure to address Irish nationalist unrest fuelled by centuries of 'landlordism'.

But changes also reflected a belief in justice and equity. As Smith had noted back in 1776, the 'dead hand' of entail and other restrictions prevented free trade in land, stopping the 'invisible hand' from allocating land to those best placed to exploit it. This free-market impulse led the great liberal philosopher J. S. Mill to establish the Land Tenure Reform Association (LTRA) in 1870. The LTRA advocated an end to primogeniture and entail, but struggled to maintain momentum after Mill's death in 1873. Somewhat more effective was the Liberty and Property Defence League (LPDL) established by the Earl of Wemyss in 1882. The LPDL saw land reform as the thin end of wedge of a socialist state tyranny. The fact that Spencer was claimed by both the LPDL and the Land Nationalisation Society

[13] Wallace, *My Life*, 2 vols (Westmead: Gregg, 1969), 1: 236.

(LNS) established by Wallace in 1881 serves as yet another reminder of the dangers of equating Spencer with a single political ideology.

Owen and Spencer formed Wallace's views on political economy. Political controversies surrounding the 'Irish Land Question' and Irish MP Charles Stuart Parnell's vision of 'peasant proprietorship' in the late 1870s nudged Wallace to get involved more directly. Another catalyst behind Wallace's decision to take a lead in the land nationalisation movement was his reading of Henry George's economic treatise, *Progress and Poverty* (1879). Born in Philadelphia, George (1839–97) was a well-travelled autodidact who had worked as a sailor, printer and prospector before settling down to work as a journalist in San Francisco. During a visit to New York George had been struck by the contrast between the relative equality of California and the stark contrasts of rich and poor visible in New York. The community which epitomized the opportunity and enterprising spirit of the United States was certainly more prosperous than California, but great wealth seemed to have brought great inequality with it, while 'labour-saving' technological innovation had done nothing to improve the lot of the working man.

'This association of poverty with progress is the great enigma of our times,' George wrote, 'so long as all the increased wealth which modern progress brings goes but to build up great fortunes, to increase luxury and make sharper the contrast between the House of Have and the House of Want, progress is not real and cannot be permanent.' The solution did not lie in the relations of 'Capital' (i.e. employers and investors) and 'Labour' (the unions); not only would 'Capital''s ability to sit out a strike always exceed that of 'Labour', such strikes ended up harming both. The root of inequality lay not in competition itself, but in the lopsided nature of the competition, itself a result of the unequal distribution of land.

The minority who controlled the vast majority of the land monopolized that unearned increment which came from the increase of the value of land in a developing economy. They became richer even as they slept, George noted, even though the increase in value (the 'rent') was 'the creation of the whole community' and so belonged to 'the whole community'. How could this rent be appropriated by the community? George proposed to do so by taxation, abolishing all taxes apart from a tax on land values. George quoted Ricardo and other political economists to the effect that a tax on rent would not reduce productivity. To offer existing landowners compensation (as Mill had argued) was to acknowledge a right to exclusive property in land which not only transcended generations (as if one generation could bind the next), but which also originated in unjust appropriation of a shared good.

George dismissed the prevailing evolutionary belief which had humanity advancing relentlessly towards a higher civilization, citing the remains of lost civilizations as well as the arrested state of other civilizations (he cited the Chinese) as evidence. Rather than seeing the advancement or regression of a civilization as a product of faculties in the individual he saw this as determined by 'social environment', which he described as 'a web of knowledge, beliefs, customs, language, tastes, institutions and laws'. It enabled acquired characteristics to be passed on, but could also be a straightjacket, preventing change. The motive force behind such change was 'mental power', whose ultimate purpose was 'the betterment of social conditions' and 'the extension of knowledge'. The more equal and associative a society, the less of this power would be pointlessly expended (on warfare, struggle, individual desires) and the more that could be devoted to human progress.

George's ideas influenced Wallace's thought on political economy, inequality, technology and Victorian progress, from his *Land Nationalisation: Its Necessity and Aims* (1882) right up to his very last book, *Social Environment and Moral Progress* (1913), which decried cities as 'the disease-products of humanity', hindering the development of humanity's higher faculties. Wallace welcomed George during his tour of the UK in 1883–84 and helped to spread his ideas during his (Wallace's) own tour of the US in 1886–87. In *Land Nationalisation* he proposed an Act of Parliament by which all land would become property of the state, which would then rent it out for an annual 'quitrent' based on the plot's assessed value. Current landowners and any would-be heirs alive at the time of the Act's passing would receive a pension equivalent to the income they had received prior to nationalization. Sub-letting would be prohibited, ensuring that nobody rented any more land than they actually needed. A similar process would see the nationalization of urban property.

The ideas of *Land Nationalisation* were shocking for their times. The rights of property owners were widely felt to be absolute, even if the distribution of land was extremely lopsided, with 536 aristocrats owning 20 per cent of the land. The idea of public access to private land for recreation, for example, was strongly resisted (e.g. by the LPDL), while the archaeologist and MP William Lubbock's 1877 attempt to secure even the most basic statutory protection for historic monuments on private land was fiercely resisted: if the owner of, say, Stonehenge wanted to knock it down and build houses on it, that was his sacred right. Wallace's first choice of publisher, Alexander Macmillan backed out of publishing it, citing threats. Whether Wallace was being socialist in advocating land nationalization, however, is far from clear.

In the 1880s his views certainly adopted a fiercer tone, linking laissez-faire capitalism and empire as conspiring together for world domination ('globalization', perhaps, though the term did not exist). *Bad Times* (1885) highlighted the destructive effects of capitalist speculators and their encouragement of militarism. Like Spencer, he made a point of criticizing the excesses of big business during his US tour. Empire was deliberately exploitative, rather than, say, a project in diffusing 'civilisation' in which the odd crack of the whip might be necessary. The combination of imperialist aggression and capitalism was noted by Marx in his *Kapital* (1867) as well as by British economists such as J. A. Hobson, in his book *Imperialism* (1902), which in turn influenced Lenin. But Wallace did not see history in terms of social classes, nor did he see capitalism as an unpleasant yet necessary step on the road to socialism. Above all, he remained an individualist, with Spencer's suspicions of the state. Once it had changed land tenure Wallace's future state had as little to do as Spencer's had. When it came to the family Wallace was traditional. Socialism would set women free to select men on grounds of their 'higher' faculties (rather than their wealth), making them 'regenerators of the race'. But monogamy and the nuclear family were not to be touched.

'I am a Socialist because I am a believer in evolution,' declared Annie Besant in her book *Why I am a Socialist* (1886). There is a sense in which similar protestations by socialists such as Marx's son-in-law Edward Aveling and Enrico Ferri were intended to lend socialism respectability and perhaps a sense of inevitability. Viewed in one way, natural selection implied 'dog-eat-dog' competition among individuals of the same species. In the case of social animals such as humans, however, evidence of cooperation and altruism could easily be found, provided one shifted the unit of selection from the individual human to the tribe or extended family group. As Darwin noted in *The Descent of Man*, a 'sympathetic' tribe whose members look out for and protect one another will outperform a tribe without mutual trust. The Russian aristocrat Prince Peter Kropotkin collected together such examples in a series of essays on 'Mutual Aid' published in *The Nineteenth Century* in the 1890s.

Though Darwin could see how 'kinship selection' (to use the modern term) worked to foster cooperation in primitive humans, he could not see how it might be applied in his own times by socialists and others attempting to secure Labour a fairer share of the benefits (in time as well as wages) of capitalist business. Just a year after *Descent* he was writing to a foreign correspondent complaining of the dysgenic effects of trade unions, noting that they insisted that all workers earn the same pay and work the same hours. Such a rule fostered unity of purpose that might be rather useful in, say, a strike. Darwin could see the employer's perspective (wanting cheap, efficient working men to outbreed the other kinds), not the worker's.[14]

Equal pay and hours featured in *Looking Backward* (1888), a work of utopian science fiction that helped Wallace imagine what his socialist future ought to look like. The hero of Edward Bellamy's fantasy goes to sleep in Boston in 1887 and awakes in the year 2000, to discover that his familiar world of fear, corruption, inequality, industrial unrest and international conflict has been totally transformed. In this brave new world men only work between the ages of 21 and 45. All trades are of equal rank and earn equal pay, distinguished only by the length of the working day (short for arduous trades, longer for the less physically demanding ones). There is no domestic service (the very word 'menial' is now archaic), thanks to machinery, while cooking is done in public kitchens. An international council administers relations between nations. There is no money, only 'credit cards'. Far from being credited with helping bring this utopia about, 'the followers of the red

---

[14]Darwin to Fick, 26 July 1872. Richard Weikart, 'A Recently Discovered Darwin Letter on Social Darwinism,' *Isis* 86 (1995): 609–11.

flag' are revealed to have been secretly subsidized by 'the great monopolies' to 'talk about burning, sacking and blowing people up, in order, by alarming the timid, to head off any real reforms'.[15]

ASSASSIN AND ALCHEMIST: ANNIE BESANT (1847–1933)

Besant's *Why I am a Socialist* acknowledged popular preconceptions of what it meant to be a socialist. 'A socialist is supposed to go about with his [*sic*] pocket full of bombs and his mind full of assassination,' she noted. For her, however, the 'very wildness of the epithets launched' indicated that non-socialists recognized that the game was up: once the working classes learned what socialism really was, 'the present system is doomed'. Besant had been drawn to socialism by an awareness that 'modern civilisation' had failed, that its prosperity produced slums as well as great wealth. The aristocracy's monopoly of land and the employer's abuse of labour-saving machinery denied the working man the just reward of his labour.

But the primary factor in persuading her had been evolution:

> The great truths that organisms are not isolated creations, but that they are all linked together as parts of one great tree of life; that the simple precedes the complex; that progress is a process of continued integrations, and ever-increasing differentiations; these truths applied to the physical animated world by Darwin, Huxley, Haeckel . . . have unravelled the tangles of existence.

Spencer had shown that 'the progress of society has been from individualistic anarchy to associated order; from universal unrestricted competition to competition regulated and restrained by law, and even to partial co-operation in lieu thereof'.[16] One cannot help wondering if she had read Spencer's *Man Versus the State*.

Besant's knowledge of evolution may have derived from attending London University, where she began a degree in science in 1879. She was tutored by Edward Aveling, a lecturer in Comparative Anatomy at the London Hospital who later married Karl Marx's daughter Eleanor. In both Aveling and Besant's cases the road to socialism led through religious doubt and free thought. Born to a middle-class Irish family, Besant married an Anglican priest. After toying with the Oxford Movement she came to reject orthodox Anglicanism, scandalizing her husband and leading to the breakdown of her marriage in 1873.

Left with a small income, Besant supported herself as a journalist and joined Charles Bradlaugh's National Secular Society, a body established in 1866 that continues to advocate the complete separation of Church and State. The pair faced obscenity charges in 1877 after publishing literature on contraception. Together with her ill-defined relationships with Bradlaugh and Aveling this made Besant seem 'unwomanly' to the vast majority of her contemporaries and confirmed a tendency

---

[15] Edward Bellamy, *Looking Backward* (London: Frederick Warne, 1891), p. 145.

[16] Annie Besant, *Why I am a Socialist* (London: Besant and Bradlaugh, 1886), pp. 2–3.

to equate socialism with the collapse of the nuclear family as the foundation of decent, 'civilized' society. As the male pronouns in the above passage suggest, Besant did not define herself by her gender.

Despite successes organizing a trade union for the match-girls at Bryant and May's London factory and as an elected member of the London School Board, after 1888 Besant became interested in Theosophy, putting Madame Blavatsky up in her London home. In 1893 Besant moved to India, where she practised theosophical chemistry and nationalist politics while awaiting the advent of the World Teacher. Together with Charles Leadbetter she undertook pioneering research into nuclear physics. *Occult Chemistry* (1908) showed the potential of the 'third eye' as a tool with which to perceive subatomic particles then beyond the reach of any microscope. Though not mainstream, the revival of alchemy that accompanied the early years of atomic research, the celebration of radium (discovered by the Curies in 1898) as 'elixir of life' and the establishment of the Alchemical Society, London (1912) by, among others, Glasgow University's Regius Professor of Chemistry (John Ferguson) make it hard to dismiss her as a lonely crank.

The view from Davos

As the *Oxford English Dictionary* (*OED*) informs us, the word 'sustainability' existed in Wallace's day, but only in a legal context: an argument either works in court (is 'sustained' by the judge) or it is 'unsustainable'. Only in the 1980s did the more familiar, 'environmental' sense appear, that is, 'sustainability' as 'the degree to which a process or enterprise is able to be maintained or continued while avoiding the long-term depletion of natural resources'. The conservation of wildlife was then seen as a question of protecting specific species from extinction at the hands of humanity. With greater understanding of climate change, however, we have come to recognize that *Homo sapiens* is itself at risk of extinction. The 1992 Convention on Biological Diversity reflected this shift from conservation as the preservation of certain charismatic animals on aesthetic grounds to an emphasis on human well-being.

The mature Wallace could be seen as an early advocate of this anthropocentric model of sustainability. When Wallace described the heavy nuts of the durian tree in *The Malaya Archipelago* he interpreted them as a lesson not to adopt a Paleyite anthropocentric model. The fact that these edible nuts were very heavy and very high up (and hence can fall and hurt men walking below) supposedly taught one not to view nature as 'organized with exclusive reference to the use and convenience of man'.[17] He later came to see humans as the goal of the second phase of evolution, that governed

[17]Wallace, *The Malay Archipelago: The Land of the Orang-utan, and the Bird of Paradise* (New York: Harper, 1869), p. 87.

by natural selection, assisted by some 'influx' to make the final jump from ape to human. With the advent of talking, thinking, tool-making humans' diversity ceased to be a question of morphology and became a question of mental culture (this is why there were no physical distinctions between races of human).

'It is when we look upon man as being here for the very purpose of developing diversity and individuality,' Wallace wrote in *The World of Life* (1910), 'to be further advanced in a future life, that we see more clearly the whole object of our earth-life as a preparation for it.'[18] Wallace's socialism was solely concerned with creating a 'social environment' in which individuality and association would be sustained and grow. Seen in this way, Wallace's socialism and spiritualism may well fit together. A concern with sustainability helps to explain his concern at the effects of human use of timber, coal and other natural resources. He was among the first to perceive how sustainability represented an unwritten compact between generations: between the living and those in the 'spirit world' as well as between the living and their immediate descendants on earth.

In 1896 Wallace was invited to speak at an international conference in Davos, Switzerland, by the Methodist and Liberal Henry Lunn (1859–1939). The British had played a crucial role in establishing the Swiss Alps as Europe's elite playground, John Tyndall being just one of the many British alpinists to popularize mountain climbing and mountain air as a health-giving escape from modern life. Starting in 1891 Lunn had organized a series of church conferences at another Swiss village, Grindelwald. Although his aim to reunify the world's protestant churches failed, he continued to work for international peace and reconciliation, notably as part of the League of Nations established after World War I. Grindelwald and Davos were perfect places to bring together leaders of different nations and faiths to think big thoughts about humanity: neutral territory, quite literally. Lunn's involvement in organizing transport to such events formed the foundation of a travel agency, Lunn-Poly, and established a tradition which continues today in the annual meetings of the World Economic Forum, also held at Davos.

An early example of 'Davos Man', Wallace had been invited to survey the nineteenth-century's achievements as a civilization. The resulting lecture became a book: *The Wonderful Century: Its Successes and Failures*. Wallace noted the technological wonders which had transformed humanity's relationship to time and space (railroads, steamships, telegraphs, undersea cables, telephones, artificial light in the shape of gas and electricity), to work (labour-saving machinery of all kinds) and health (anaesthetics, germ theory, antiseptics, X-rays, but not vaccination). He discussed great strides in physics, chemistry, astronomy, geology and biology, briefly describing his own discovery of natural selection. He compared the century's achievements and discoveries against all previous centuries. None came close to matching

[18]Wallace, *The World of Life* (London: Chapman and Hall, 1914), p. 397.

it. 'Both as regards the number and the quality of its onward advances, the age in which we live fully merits the title I have ventured to give it of – THE WONDERFUL CENTURY.'[19]

But Wallace does not end there, on page 156. He goes on to consider the 'Failures', which take up more pages (222 of them) than the 'Successes'. This is, admittedly, partly thanks to a long chapter opposing vaccination, which Wallace cut from later editions. The other chapters, on 'The Demon of Greed', 'Militarism – the Curse of Civilization' and 'The Plunder of the Earth' nonetheless make for grim reading, grimmer now, perhaps, than in 1898. Less compelling is Wallace's insistence that his century's neglect of, and opposition to phrenology, spiritualism and psychical research will be redressed by the next. Though Wallace could see how the British Empire might turn out to be for the good of the human race, he concluded that 'we deserve no credit for it', British rule having been largely influenced by 'the necessity of finding well-paid places for the less wealthy members of our aristocracy'.[20]

Wallace reached his climax in his account of how 'the struggle for wealth' had robbed the living and the unborn

> by a reckless destruction of the stored-up products of nature, which is even more deplorable because more irretrievable. Not only have forest-growths of many hundreds of years been cleared away, often with disastrous consequences, but the whole of the mineral treasures of the earth's surface, the slow products of long past eons of time and geological change, have been and are still being exhausted, to an extent never before approached, and probably not equalled in amount during the whole preceding period of human history.[21]

Wallace was not the first to note the unsustainable rate at which Britain's coal reserves were being depleted. That honour belonged to the economist Walter Stanley Jevons, who published *The Coal Question* in 1865. Wallace had listed the too-rapid extraction among the sins of aristocratic landlordism critiqued in *Land Nationalisation*. Coal had become, like water and land itself, a resource essential to human life. It followed that private ownership of such resources was unjust.

Nonetheless Wallace was the first to relate 'the coal question' to a broader critique of the unsustainable exploitation of natural resources, and to note the effect of deforestation on diversity. *Tropical Nature* (1878) noted how farmers' attempts to extract the maximum profit from, say, coffee growing led them to clear-cut jungle rather than leave belts of trees that protected

---

[19]Wallace, *The Wonderful Century: Its Successes and Failures* (New York: Dodd, Mead, 1898) p. 156.
[20]Ibid., p. 375.
[21]Ibid., p. 494.

the soil from erosion. Without such belts the fertile soil soon washed away, forcing the farmer to move on and clear-cut another patch of jungle. By 1910 he was advocating jungle reserves as a backstop to protect the unusually rich biodiversity of the rainforest, both for the 'enjoyment and study of future generations' and 'to prevent further deterioration of climate and destruction of the fertility of the soil'.[22]

Wallace had no problem with Spencer's 'survival of the fittest'; he preferred it, in fact, to 'natural selection'. But he did not see competition as a zero-sum game. If anything, evolution offered a way of escaping it. To take Darwin's finches as an example, natural selection involved finches of the same species competing for hard seeds, in a process which would result in the best-of-all-possible-beaks for manipulating and breaking open that specific variety of seed. To this extent, evolution is a zero-sum game: at any one time the finches are competing for a share of a fixed amount of seed. In the long run, of course, their activities may increase the amount of seed, say, by carrying seeds to new ground suitable for that seed-producing plant. But that is no comfort to the finches today.

But that isn't all that happens, as we know. Some finches that would otherwise lose this particular battle begin to eat, say, berries, or insects. New beaks are formed, diversity is increased. An island which was once able to sustain one species now sustains three, or four or five species, none of which are competing with one another for food. Seen in this way, evolution is about finding the best (as in, greatest-diversity-supporting) use of the available resources. Human technology and tools are part of this process. Fire allows cooking, which makes meat safer to eat and longer lasting. It also makes certain things edible that are otherwise inedible. The meat and the fruit haven't changed in their makeup or amount: the way they are used has, however.

Wallace wasn't in any doubt that the development of capitalism, technology or the empire involved an evolutionary process, that is, one full of failed experiments and unintended consequences. The question was, could their continued operation under the guise of 'growth', 'advancement' or 'civilisation' be justified in universal terms, as part of a 'spirit world' which did not just maintain current levels of diversity, but increased them? Here the word 'sustainability' fails, with its tragic insistence that we draw a line around what we have now, and try to slow down inevitable loss as much as possible. Things cannot get better. Evolution has stopped working. Things can only get worse. Wallace's 'sustainability' was much more hopeful, pushing frontiers and claiming new territory. To that extent Wallace, for all his contrariness, remains a Victorian.

[22]Wallace, *The World of Life*, p. 77.

Further reading

Barr, Alan. P. (ed.), *Thomas Henry Huxley's Place in Science and Letters: Centenary Essays* (Atlanta: University of Georgia Press, 1997).

Crook, Paul, *Darwinism, War and History: The Debate on the Biology of War from 'The Origin of Species' to the First World War* (Cambridge: Cambridge University Press, 1994).

Desmond, Adrian, *Huxley. Evolution's High Priest* (London: Michael Joseph, 1997).

Fichman, Martin, *An Elusive Victorian: The Evolution of Alfred Russel Wallace* (Chicago: Chicago University Press, 2004).

Lightman, Bernard, *The Origins of Agnosticism: Victorian Unbelief and the Limits of Knowledge* (Baltimore: Johns Hopkins University Press, 1987).

—, 'Huxley and Scientific Agnosticism: The Strange History of a Failed Rhetorical Strategy,' *British Journal of the History of Science* 35 (2002): 271–89.

Oppenheim, Janet, *The Other World: Spiritualism and Psychical Research, 1850–1914* (Cambridge: Cambridge University Press, 1985).

Raby, Peter, *Alfred Russel Wallace: A Life* (Princeton: Princeton University Press, 2001).

Smith, Charles H. and George Beccaloni (eds), *Natural Selection and Beyond: The Intellectual Legacy of Alfred Russel Wallace* (Oxford: Oxford University Press, 2008).

Stack, David, *The First Darwinian Left: Socialism and Darwinism, 1859–1914* (Cheltenham: New Clarion Press, 2003).

Online Resources

The Alfred Russel Wallage Page. http://people.wku.edu/charles.smith/index1.htm
The Huxley File. http://aleph0.clarku.edu/huxley/

Conclusion: The Longest Discovery

In the century following Wallace's death scientists have further advanced our understanding of evolution, most importantly by genetics. In the 1880s the German biologist Theodor Boveri worked out that genetic information was contained within chromosomes, a structure made of deoxyribonucleic acid (DNA) and protein found in cells. It was unclear which did the work, however, until the 1940s. In 1953 James D. Watson and Francis Crick established the double-helix (twisted ladder) structure of DNA. The rungs of the ladder were made up of pairs of molecules called nucleotides. There were four kinds of nucleotides, each of which formed 'base pairs' with one of the others. Adenine pairs only with thymine, for example. Genetic information was thus 'coded' in a four-letter alphabet, much as a computer stores information in zeroes and ones. In reproduction one side of the ladder peeled off and was copied.

In 1976 Richard Dawkins advanced the notion of 'kinship selection', as a means of explaining the altruistic behaviour which Darwin and others had noted the previous century. A life form is altruistic when its actions benefit relatives, at some cost to its own survival. Sacrificing oneself to benefit one's offspring made obvious sense under natural selection. By proposing that the gene was the basic unit of selection Dawkins proposed that relatives who were *not* parents might nonetheless display altruism, because those altruists (those making the sacrifice) had much of the same genetic material as the beneficiaries did. Altruism was actually a case of selfish genes looking out for themselves. A child is 100 times more likely to be killed by a step-parent than a parent, something which would seem to indicate just how 'selfish' such gene-focused instincts can be.[1] The 'wicked stepmother' is not a fairy tale figure. She is simply displaying good evolutionary tactics.

The story does not end with Dawkins. Though we have transcribed ('sequenced') the DNA of humans, many of the promises made by the

---

[1]Michael Ruse, *Darwinism and Its Discontents* (Cambridge: Cambridge University Press, 2006), pp. 191–2.

Human Genome Project when it started in 1990 have failed to materialize. Hopes of gene therapy were based on the assumption that the genome could be neatly chopped up into pieces that controlled this or that trait. It would allow intervention to 'repair' genes by removing and replacing sequences which left the individual prone to develop a particular illness. The rise of 'evo devo' (see box) has indicated that the expression of genes is more influenced by environmental factors than we previously thought, inspiring a revival of Lamarckian evolution, in which the phenotype influences the genotype, criss-crossing the 'Weismann barrier'.

The sheer amount of information contained in our DNA and the amount of it which we share with completely different organisms, including plants, is bewildering. Why do humans have only 46 chromosomes, fewer than guinea pigs (64), dogs (78) and ferns (1,200, approximately)? Time and again large sections of genetic code have been dismissed as 'garbage', and then re-evaluated. And that is just DNA. The mitochondria inside our cells, which help us release the chemical energy in food, have their own mitochondrial DNA (mtDNA). This is a separate line of inheritance from our DNA and is believed to derive from bacteria which early forms of cellular life absorbed and enslaved a very long time ago. Then there is the single-stranded RNA, which, it has been proposed, was used by viruses (which consist of genetic material without a cell) to encode genetic material before they switched to using DNA.

Work on a 'RNA world thesis' is leading us back to Darwin's other big unanswered question: the origin, not of species, but of life itself. The bottom end of Darwin's evolutionary tree diagram may not end in a single trunk, may not emerge from a single, Lightfoot-like event. It seems increasingly likely that it ends in a tangled skein of roots, as various self-replicating chemicals forged a symbiotic or mutually beneficial relationship with one another. Exploration of this field isn't just leading us to new answers to the question 'How did we evolve?,' it is radically changing the subject of that question. Who are 'we'? *Homo sapiens*, a genome, or a cluster of genomes? The human body contains far more bacterial DNA than human DNA, and these bacteria have been shown to be critical to our body's normal function. To state that much work remains to be done is evident. There is much to mystify us.

WORK IN PROGRESS: EVOLUTION AND DEVELOPMENT

One of the assumptions made in early twentieth-century genetics was that more complex organisms, those with a larger number of specialized organs, would have more genetic material than simpler ones. The instruction manual for making a human would surely be a lot longer than that for making a nematode worm. When the Human Genome Project started to sequence human DNA in 1990, therefore, it was predicted that they would find 60,000 or more genes, many more than the 16,000 that had been found in the lowly nematode worm. It was surprising, and perhaps humbling, to find there were 'only' 25,000 genes. Complexity or simplicity was not, it seemed, a function of how much genetic material one had, but rather over how the expression of that material was coordinated.

In 1961 Jacques Monod and François Jacob found evidence that the bacterium *Escherichia coli* (a parasite that lives in mammal intestines) was 'switching on and off' parts of its genome, which consisted of around 5,000 genes. The sequence of genes affected was that responsible for producing an enzyme that helps the bacterium digest a certain sugar, lactose. According to the Modern Consensus, this sequence of genetic material should have been expressed in every member of the species. Instead Monod and Jacob found that individual bacteria were using a 'repressor molecule' to switch off this sequence of genes when there was no lactose present in their environment. While the environment was not altering the genotype (the DNA sequence was not changed), it was changing which parts of it were being expressed.

In 1995 Christiane Nüsslein-Volhard and Eric F. Wieschaus won a Nobel Prize for their work on Hox genes, a group of genes which organize developing embryos into distinct segments (head, thorax, etc.) and which produce proteins that switch on and off entire sequences of neighbouring genes (e.g. sequences responsible for making eyes). A gene shown to be linked to the formation of eyes in *Drosophila melanogaster* (fruit fly) also turned on genes that make eyes in mice. When the transfer was attempted the other way, from mouse to fly, the outcome was the same. Yet insects have compound eyes, made up of many different tubes (each with its lens and retina), while mammals like mice and humans have eyes consisting of a single lens focusing the image on a retina. It had been assumed that each model evolved separately, producing different genetic sequences which would bear no relation to one another.

This and other work on the embryological development of the fruit fly has complemented earlier work on environment and gene expression, creating a new field called 'evolution and development', or 'evo devo'. This work is suggesting that the genome is not simply an instruction manual, a DNA script containing the 'code' required to reproduce all the tiny beneficial mutations that a single organism has picked up along its lonely evolutionary path. Rather than throwing away 'code' for traits found to be harmful or simply redundant, genetic material has been retained over vast periods of time, only a fraction of which has been expressed by this or that organism at this or that point in its history. This would explain why so much of the same apparently useless genetic material is found in creatures placed on very different parts of the evolutionary tree.

It may be time to change metaphors. Our genetic material is not a unique code which defines our essence. Instead it might be seen as an antique piano, a vast keyboard shared by all forms of life. Though the keys are shared, and passed down, not all of them are played. The piano is the same, only the tune changes.

The unsettling and sometimes surprising findings of 'evo devo' need not lessen our respect for Darwin, Wallace and other Victorian men of science. But they do emphasize the fact that these Victorians are a chapter in an ongoing story of humanity's coming to terms with evolution. This book has indicated the limitations of our usual way of understanding how knowledge is advanced: as a series of moments in which scholar-recluses emerge blinking into the grateful glare of public acceptance. Evolution did not spring ready-made from Darwin's mind, nor was the debate restricted to those who fit our twenty-first-century definition of 'scientist'. The very word 'scientist' was not commonly used until the 1880s, which is why this book has used the phrase 'men of science'. Institutions, patronage, publications and public opinion – in a word, politics – played an important role in the nineteenth-century development of the theory. This book has considered science as an integral part of Victorian culture and 'the Victorian mind'. Culture conditions even the most open and scientific of minds. Indeed, Darwin's trip to the gin palace suggests that his 'genius' lay in his very openness to cultures – breeders' culture, pub culture – to which fellow members of the elite were closed. One cannot help but wonder if today's highly specialized scientists are anything like as open. That interdisciplinary conversation which a Victorian elite pursued at meetings of the Metaphysical Society and then in the pages of the periodical *The Nineteenth Century* barely survived its liveliest contributor, Thomas Henry Huxley. 'There is one cause in action over the whole field of knowledge,' Gladstone wrote to Henry Acland in 1883, 'which has a powerful tendency to reduce [humanity's] dimensions. I mean that we are all coming to be specialists.'[2]

Neither universal secondary education, a massive expansion of higher education nor the advent of the internet have managed to revive the polymath. The division between the sciences and the humanities, between the 'two cultures' identified by C. P. Snow in 1956 remains deep, and the level of scientific literacy encountered in the public at large is low. Far from bridging them, the media often fosters misunderstanding, making claims for what this or that 'recent study' supposedly 'proves' (failing to mention other studies, which may *suggest* a different conclusion) that the scientists concern can only blush (or groan) at. We still look to scientists for straightforward yes-or-no answers to problems, something they rarely are in a position to provide, rather than looking to them for the information and understanding to make our own decisions in cases where there is less certainty than we might wish (which tends to be most of the time).

Although Spencer and Wallace are important exceptions, otherwise it is fair to say that Victorians had a habit of assuming that 'evolution' and 'civilisation' (*their* civilization) went hand in hand. For them 'empire' and 'free trade' went hand in hand, too, as part of the same progressive mechanism, working to make the world a (supposedly) better place. We seem to have

---

[2]Gladstone to Acland, 29 October 1883. Bodleian Library, Oxford. MS Acland.d.68, f. 66.

weaned ourselves off that habit, encouraged by evidence of the damage that has been done to our planet by unsustainable exploitation of its finite resources, as well as to our human nature, by the selfish indifference which too often accompanies capitalist growth (at least when it is pursued as an end in itself).

We are thereby enabled to ponder questions which most Victorians only just began to formulate. Whose interests does science serve? What is the role of the State and public opinion in directing research? What are evolution's implications for us, as beings who are, at one and the same time, members of a particular world order, a species, a gender, a society and a family? Lest we begin to consider ourselves more tolerant or open-minded than those 'hidebound' Victorians, however, we also need to recognize that certain questions remain out of bounds to science. Evolutionary hypotheses can still strike us as 'offensive', as beyond discussion.

OFFENSIVE SCIENCE

In October 2007 a man named James Watson travelled to Britain to give a series of lectures on science, including at the Dana Center, part of London's Science Museum. At the start of his visit he gave an interview to the *Times* in which he claimed to be 'gloomy' about the prospects of Africa. 'All our social policies are based on the fact that [Africans'] intelligence is the same as ours,' Watson stated, 'whereas all the testing says not really.' The result was a firestorm of outrage from journalists, scientists, politicians and others.

Many agreed with the MP Keith Vaz that Watson's comments were 'deeply offensive', 'a reminder of the attitudes which can still exist at the highest professional levels'. The Commission of Racial Equality announced that it was to investigate, while the 1990 Trust called for legal action to be taken against Watson for encouraging 'bigotry'. Although the Dana Center had been established to help the Science Museum attract adult visitors by tackling exciting and controversial issues in science, it decided to disinvite Watson. Watson had passed 'beyond the point of acceptable debate'.

Belief in human equality is one of the core beliefs enshrined in British life. It is enshrined in UK laws prohibiting discrimination on ground of race, in the European and the UN Declarations of Human Rights. Laws guaranteeing free speech do not protect speech that encourages hatred towards others on grounds of race. Yet Watson was not a far-right politician, but a Nobel Prize-winning scientist with 19 honorary degrees and a knighthood. His theory was couched in terms that the co-discoverer of the structure of DNA presumably understood well. 'There is no firm reason to anticipate that the intellectual capacities of peoples geographically separated in their evolution should prove to have evolved identically.'[3]

---

[3]Charlotte Hunt-Grubbe, 'The Elementary DNA of Dr Watson,' *The Times*, 14 October 2007.

As a working, scientific hypothesis, this is sound enough. To propose it as a hypothesis 'is not to give in to racism', Watson concluded. Princeton bioethicist Peter Singer agreed. Singer went on to note that 'Whatever science uncovers in the field of race and intelligence' the findings 'will not justify racial hatred, nor disrespect for people of a different race. Whether some are of higher or lower intelligence has nothing to do with that.'[4] The Dana Center could have pointed to scientific studies indicating, say, that IQ tests are culturally biased. It could have invited a second scientist to debate with Watson, citing the many studies which disagree with his hypothesis. Instead it concluded that his hypothesis was 'unacceptable'.

Belief in God and the doctrines of the Church of England was enshrined in nineteenth-century British life. It was enshrined in laws prohibiting members of other churches as well as atheists from sitting as MPs, attending Oxford and Cambridge and doing much else besides. Laws guaranteeing free speech did not protect speech that encouraged disrespect towards God in others (blasphemy). Admittedly, over the course of the century many of these civil disabilities were removed or allowed to fall into disuse. But the simple truth remains, that evolutionary theories associated with materialism and atheism were deeply offensive to many Victorians. Hence Darwin's 20-year silence, and the shocked response of his contemporaries.

In 1860 another scientific institution – the British Association for the Advancement of Science – had been faced with the choice of whether to allow the discussion of an offensive new theory within the Natural History Museum in Oxford. It went ahead. Darwin's theory should be heard and tested against established theories, even if it seemed to contain (as Darwin put it) 'much odious truth'. As dwellers in the twenty-first century, we might be tempted to conclude the Victorians' feverish response to evolution was exaggerated, the product of a hidebound society. The case of Prof Watson, however, suggests otherwise.

Although Darwin kept his own speculations on the political or social implications of evolutionary mechanisms for humanity private, he certainly did not discourage others from speculating. As we have seen, evolution by natural selection was claimed by both imperialists and anti-imperialists, and by conservatives, liberals, socialists and anarchists. This book has challenged the tendency to emphasize the association with the first, that is, to focus on Social Darwinism, a term which was coined halfway through the twentieth century with a clear agenda in mind. By now the reader is in a position to evaluate the utility (or not) of this term, in the light of a fuller consideration

[4]Peter Singer, 'Should We Talk About Race and Intelligence?', Project Syndicate, 1 November 2007. http://www.project-syndicate.org/commentary/should-we-talk-about-race-and-intelligence- (accessed 2 May 2013).

of Spencer than is usually found in books on Victorian evolution. The socialist Darwinism of Alfred Russel Wallace is equally important to such a discussion. Rather than seeing Wallace's socialist (and spiritualist) activity as representing distinct phases in a career supposedly scarred by inconsistency, this book has indicated the continuities in his thought.

Identifying a point at which such 'uses' of evolutionary science become 'abuses' is difficult. The approach taken here has been to challenge the question itself, by refusing to identify a clear separation between the institutions in which 'the science is done' and the rest of the world, where it is 'consumed'. Far from being dispassionate, science is an activity driven by human passions: curiosity, ambition and benevolence, but also resentment and greed. That fact is sometimes overlooked or even denied, as it arguably was in the closing years of the nineteenth century. At such times science becomes alien, a threat to humanity from 'outside,' rather than another expression of it. Humanity can only benefit from an open, informed and patient public conversation about the latest discoveries in evolution and their implications for Huxley's 'question of questions', that question posed in the introduction to this book. As historians and as humans, we surely have something to contribute to this project.

Further reading

Carroll, Sean B., *Endless Forms Most Beautiful: The New Science of Evo Devo and the Making of the Animal Kingdom* (New York: Norton, 2006).

Dawkins, Richard, *The Selfish Gene* (Oxford: Oxford University Press, 1976).

Lewontin, R. C., *The Doctrine of DNA: Biology as Ideology* (New York: Harper, 1993).

Ruse, Michael and William A. Dembski, *Debating Design. From Darwin to DNA* (Cambridge: Cambridge University Press, 2004).

GLOSSARY

'What a long word!'
'If you can find a shorter one I shall be very much obliged to you, for I hate long words.'

Kingsley, 'The World's End',
Madam How and Lady Why (1870)

As historians we need to be especially sensitive to the words we use: to ensure precision, avoid anachronism and above all to make sure that we are getting the most out of the written sources we consult – that we are 'reading it their way'. The following glossary defines some key technical terms commonly used in the Victorian period, as well as some other words that have changed their meaning in the intervening period. The date in brackets indicates the first recorded usage in English of this particular sense of the word. A word may well have existed earlier, without being used in the sense Victorian men of science used it. For words not included here the reader is directed to the *Oxford English Dictionary*, a wonderful resource which allows the user to trace the different ways in which words have been used over the years, by reading the short quotations following each definition.

Agnostic, n., A person who believes it to be impossible for humans to establish the basis of knowledge or gain knowledge of immaterial things, to establish the existence or nature of God (Huxley 1869). Contrasted with an atheist, for whom the non-existence of God is a matter of conviction and certainty.

Altruism, n., Devotion to the welfare of others, as a principle of action (1853 – originally coined by the French philosopher Comte).

Ammonite, n., A fossil genus of cephalopods, consisted of spiral-shaped shells (1758).

Anthropology, n., The science of mankind, in the widest sense. From about 1860 it started being used in the more familiar sense of the branch of science which investigates mankind zoologically, often in racial terms.

Archaeopteryx, n., A fossil bird-like creature with a long vertebrate tail (1859).

Archetype, n., A key concept in comparative anatomy; the ideal model or pattern assigned to each division of living creatures (1849).

Biogeography, n., The science which considers the distribution of living things across space.

Biology, n., Originally used to denote the study of human character (1813). Only later used to refer to the division of physical science which deals with organized beings or animals and plants (Whewell 1847).

Catastrophist, adj., Holding with the theory by which the earth's history is understood to consist of long periods of stasis punctuated by sharp periods of massive volcano, earthquake and flood activity. Opposite of *uniformitarian*.

Darwinistic, adj., Used by historians of science to describe caricatured or crude applications of Darwin's theory, for example in *social darwinism*.

Degradationism, n., The belief that organisms become simpler or cruder, implying the reversal of progressive development.

Development, n., Darwin preferred to use 'development' or 'the developmental hypothesis' to describe what we call evolution, perhaps following *Vestiges*' use of it to refer to 'the evolutionary process and its result'.

Disestablishment, n., The process by which the Anglican Church (Church of England) lost its special status and privileges as the State Church. Linked with the removal of restrictions which previously prevented non-Anglicans (Roman Catholics, Dissenters, Jews) from sitting in Parliament, attending certain universities, etc.

Dissenter, n., A member of a non-Anglican Protestant Church (e.g. Baptists, Methodists), also known as a Nonconformist.

Dysgenic, adj., Exerting a detrimental effect on the race, tending towards racial degeneration. (1915). The opposite of *eugenic*.

Empiricism, n. The philosophical theory that all knowledge is derived from the experience of our senses.

Eugenics, n., The science which attempts to engineer the production of fine offspring (Galton 1883). The opposite of *dysgenic*.

Evolution, n., Literally 'unfolding'. The process by which an organism develops from immaturity to maturity (1670). We use it in the sense first used by Lyell in 1832: 'The origination of species of animals or plants, as conceived by those who attribute it to a process of development from earlier forms.'

Homologous, adj., Having the same relation, proportion or relative position – corresponding in type of structure. Said of parts or organs in different animals or plants (Owen 1846). These similarities were called *homologies*.

Laissez-faire, n., The principle that government should not interfere with the action of individuals, esp. in trade (1825).

Materialism, n., The theory or belief that nothing exists except matter and its movements or modification. Often associated with atheism.

Metaphysics, n., The branch of philosophy which addresses fundamental concepts such as knowing and being.

Monogenist, adj., A person who believes that all humans descend from a single pair of ancestors (1857). The opposite of *polygenist*.

Morphology, n., The branch of *biology* that deals with the form of living organisms and their parts, and the relationships between their structures (teeth, bones, organs).

Mutation, n., An abrupt transition producing an organism with heritable characteristics (not necessarily *dysgenic*) differing markedly from those of the parent. (1901).

Natural Selection, n., The evolutionary theory of Darwin, of the preferential survival and reproduction of organisms better adapted to their environment.

Natural Theology, n., The study of the created world (nature) as a means of learning about the Creator (God).

Naturphilosophie, n., The late eighteenth-, early nineteenth-century German system of zoology which understood life as the expression of a transcending Absolute, realized in certain basic, ideal forms (somewhat like *archetypes*).

Nebular Hypothesis, n., The theory that galaxies emerged from a process by which a nebula (cloud of gas) collapsed into solid matter, into a rotating galaxy made up of planets orbiting suns.

Ontogeny, n., The origin and development of an individual organism (1872). See *phylogeny*.

Pantheism, n., The belief that the universe or nature is a manifestation of God. Widely condemned in the Victorian period as denying the personality of God (as a set of traits) and Jesus Christ, as a faith without a religion.

Persistence, n., Continued unchanging existence (1849). Often used in contrast to *saltation*, to describe a long period in which creatures did not mutate.

Phenotype, n., The characteristics of an individual life form understood as the result of the interaction of the *genotype* with the life form's environment (1910).

Phrenology, n., The theory that personality consists of separate faculties, each located in an organ found in a definite region of the surface of the brain, so as to enable the 'bumps' to be 'read' for evidence of individual character.

Polygenist, n., A person who believes that different races of mankind arose independently of one another (1857).

Positivism, n., A philosophical system developed from the 1830s by the French thinker Auguste Comte (1798–1857), one which recognizes only observable phenomena and testable laws, which rejects inquiry into ultimate causes.

Providence, n., The foreknowing and protective care of God – divine direction, control or guidance (1382).

Recapitulation, n., The repetition of evolutionary stages in a young animal (1875) – usually while still an embryo/foetus.

Revealed, adj., Brought to light or disclosed by divine or other supernatural agency (1562). Hence revelation, knowledge gained directly from God.

Saltationism, n. (Latin 'saltus' = a leap) A model of evolutionary development organized around sharp changes or *mutations*, especially ones with marked effects on several aspects of an organism (Huxley 1870). See *persistence*.

Scientism, n., Excessive belief in the power of scientific techniques.

Scientist, n., A person with expert knowledge of a science, a qualified specialist (as opposed to layman) using scientific methods (1834, but not widely used until the 1870s).

Sexual selection, n., Selection which arises through the preference by one gender (usually the female) for those individuals of the other gender that have some special characteristic indicative of superior fitness.

Social Darwinism, n., The application (or misapplication) of evolutionary concepts of competition in public policy concerning trade or public health (1890s). Often associated with *laissez-faire* economics.

Special Creation, n., A 'one off' act of divine creation, resulting in a specific organism rather than a vast array of life forms.

Speciation, n., The formation of new species in the course of evolution (1906).

Spontaneous Generation, n., The sudden appearance of a life form without any assistance from parent life forms, for example, by means of passing an electric current through water or a chemical bath. A highly controversial theory Victorians associated with *materialism*.

Sport, n., A mutant offspring, a freak of nature.

Superfecundity, n., The tendency of life forms to increase their numbers exponentially.

Taxonomy, n., The systematic classification of living organisms (1813) into ranks (species, genus, etc.) or taxa (singular: taxon).

Teleology, n., (Greek 'telos' = end) The understanding of phenomena (e.g. *development*, human history) as guided towards a certain end or purpose, rather than following a random 'course.'

Theism, n., Belief in God as creator and ruler of the universe, but not, say, as father of Jesus Christ or a personified deity.

Transmutation, n.,Transformation of one species into another (1859).

Uniformitarianism, n., The theory that geological processes have always been driven by causes observable today (Whewell 1840). Opposite of *catastrophist*.

Vestigial Organ, n., A shrunken or undeveloped organ; one which has ceased to perform a useful function many generations ago. The identification of such organs provided evidence for evolution.

INDEX